Cold Atmospheric Plasma (CAP) Technology and Applications

Synthesis Lectures on Mechanical Engineering

Synthesis Lectures on Mechanical Engineering series publishes 60–150 page publications pertaining to this diverse discipline of mechanical engineering. The series presents Lectures written for an audience of researchers, industry engineers, undergraduate and graduate students. Additional Synthesis series will be developed covering key areas within mechanical engineering.

Cold Atmospheric Plasma (CAP) Technology and Applications
Zhitong Chen and Richard E. Wirz
July 2021

Capstone Engineering Design: Project Process and Reviews (Student Engineering Design Workbook)
Ramana Pidaparti
June 2021

Asymptotic Modal Analysis of Structural and Acoustical Systems
Shung H. Sung, Dean R. Culver, Donald J. Nefske, and Earl H. Dowell
November 2020

Machine Design for Technology Students: A Systems Engineering Approach
Anthony D'Angelo Jr.
October 2020

The Engineering Dynamics Course Companion, Part 2: Rigid Bodies: Kinematics and Kinetics
Edward Diehl
September 2020

The Engineering Dynamics Course Companion, Part 1: Particles: Kinematics and Kinetics
Edward Diehl
September 2020

Fluid Mechanics Experiments
Robabeh Jazaei
September 2020

Sequential Bifurcation Trees to Chaos in Nonlinear Time-Delay Systems
Siyuan Xing and Albert C.J. Luo
September 2020

Introduction to Deep Learning for Engineers: Using Python and Google Cloud Platform s
Tariq M. Arif
July 2020

Towards Analytical Chaotic Evolutions in Brusselators
Albert C.J. Luo and Siyu Guo
May 2020

Modeling and Simulation of Nanofluid Flow Problems
Snehashish Chakraverty and Uddhaba Biswal
March 2020

Modeling and Simulation of Mechatronic Systems using Simscape
Shuvra Das
March 2020

Automatic Flight Control Systems
Mohammad Sadraey
February 2020

Bifurcation Dynamics of a Damped Parametric Pendulum
Yu Guo and Albert C.J. Luo
December 2019

Reliability-Based Mechanical Design, Volume 2: Component under Cyclic Load and Dimension
Design with Required Reliability
Xiaobin Le
October 2019

Reliability-Based Mechanical Design, Volume 1: Component under Static Load
Xiaobin Le
October 2019

Solving Practical Engineering Mechanics Problems: Advanced Kinetics
Sayavur I. Bakhtiyarov
June 2019

Natural Corrosion Inhibitors
Shima Ghanavati Nasab, Mehdi Javaheran Yazd, Abolfazl Semnani, Homa Kahkesh, Navid Rabiee, Mohammad Rabiee, and Mojtaba Bagherzadeh
May 2019

Fractional Calculus with its Applications in Engineering and Technology
Yi Yang and Haiyan Henry Zhang
March 2019

Essential Engineering Thermodynamics: A Student's Guide
Yumin Zhang
September 2018

Engineering Dynamics
Cho W.S. To
July 2018

Solving Practical Engineering Problems in Engineering Mechanics: Dynamics
Sayavur Bakhtiyarov
May 2018

Solving Practical Engineering Mechanics Problems: Kinematics
Sayavur I. Bakhtiyarov
April 2018

C Programming and Numerical Analysis: An Introduction
Seiichi Nomura
March 2018

Mathematical Magnetohydrodynamics
Nikolas Xiros
January 2018

Design Engineering Journey
Ramana M. Pidaparti
January 2018

Introduction to Kinematics and Dynamics of Machinery
Cho W. S. To
December 2017

Microcontroller Education: Do it Yourself, Reinvent the Wheel, Code to Learn
Dimosthenis E. Bolanakis
November 2017

Solving Practical Engineering Mechanics Problems: Statics
Sayavur I. Bakhtiyarov
October 2017

Unmanned Aircraft Design: A Review of Fundamentals
Mohammad Sadraey
September 2017

Introduction to Refrigeration and Air Conditioning Systems: Theory and Applications s
Allan Kirkpatrick
September 2017

Resistance Spot Welding: Fundamentals and Applications for the Automotive Industry
Menachem Kimchi and David H. Phillips
September 2017

MEMS Barometers Toward Vertical Position Detection: Background Theory, System Prototyping, and Measurement Analysis
Dimosthenis E. Bolanakis
May 2017

Engineering Finite Element Analysis
Ramana M. Pidaparti
May 2017

Cold Atmospheric Plasma (CAP) Technology and Applications
Zhitong Chen and Richard E. Wirz

ISBN: 978-3-031-79700-2 print
ISBN: 978-3-031-79701-9 ebook
ISBN: 978-3-031-79702-6 hardcover

DOI 10.10007/978-3-031-79702-6

A Publication in the Springer series
SYNTHESIS LECTURES ON MECHANICAL ENGINEERING
Lecture #35

Series ISSN 2573-3168 Print   2573-3176 Electronic

Cold Atmospheric Plasma (CAP) Technology and Applications

Zhitong Chen
National Innovation Center for Advanced Medical Devices, Shenzhen, China

Richard E. Wirz
Mechanical and Aerospace Engineering, University of California, Los Angeles, California, U.S.A.

SYNTHESIS LECTURES ON MECHANICAL ENGINEERING #35

ABSTRACT

Cold atmospheric plasma (CAP) is a promising and rapidly emerging technology for a wide range of applications, from daily life to industry. CAP's key advantage is its unique ability to effectively deliver reactive species to subjects including biological materials, liquid media, aerosols, and manufactured surfaces. This book assesses the state-of-art in CAP research and implementation for applications including agriculture, medicine, environment, materials, catalysis, and energy. The mechanisms of generation and transport of the key reactive species in the plasma are introduced and examined in the context of their applications. Opportunities and challenges for novel technologies, fresh ideas/concepts, expanded multidisciplinary study, and new applications are discussed. The authors' vision for the converging trends across diverse disciplines is proposed to stimulate critical discussions, research directions, and collaborations.

KEYWORDS

cold atmospheric plasma, plasma agriculture, plasma medicine, plasma materials, plasma environment, plasma catalysis, plasma energy

Contents

Acknowledgments xiii

1   Introduction   1

2   Cold Atmospheric Plasma (CAP)   7
   2.1   CAP Sources   7
   2.2   CAP Diagnostics   9
   2.3   CAP Simulation and Modeling   13
   2.4   Summary and Future Directions   21

3   Plasma Agriculture   23
   3.1   Seeds   23
   3.2   Crops   28
   3.3   Foods   30
   3.4   Summary and Future Directions   32

4   Plasma Medicine   35
   4.1   Disinfection   35
   4.2   Wound Healing   39
   4.3   Cancer Therapy   43
   4.4   Other Biomedical Applications   53
   4.5   Summary and Future Directions   56

5   Plasma Environment   59
   5.1   Air Pollution   59
   5.2   Water Pollution   63
   5.3   Soil Pollution   66
   5.4   Summary and Future Directions   69

6   Plasma Materials   71
   6.1   Surface Modification   72
   6.2   Material Synthesis   77
   6.3   Summary and Future Directions   81

7   Plasma Catalysis **83**

    7.1   Plasma Treatment of Catalysts 84

    7.2   PDC and PAC Reactions 86

    7.3   Other Catalysis Factors 89

    7.4   Summary and Future Directions 93

8   Plasma Energy **95**

    8.1   Chemical Storage 95

    8.2   Direct Electrical Storage/Production 97

    8.3   Summary and Future Directions 101

9   Conclusions, Challenges, and Future Work **103**

    9.1   Overview/Conclusion 103

    9.2   CAP Fundamental Research (Across Applications and
          Cross-Disciplinary) 106

References **109**

Authors' Biographies **191**

Acknowledgments

The authors acknowledge many people who contributed to this book, both directly and indirectly, but especially Richard Obenchain for his valuable contributions. This work was supported by grants from the start-up packages of the National Innovation Center for Advanced Medical Devices (NMED) and the Air Force Office of Scientific Research (FA9550-14-10317, UCLA Subaward No. 60796566-114411).

CHAPTER 1

Introduction

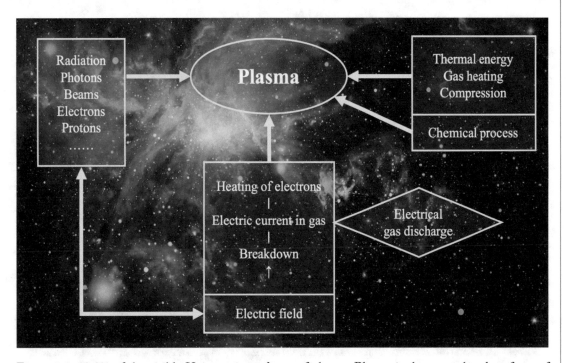

Figure 1.1: 99.9% of the visible Universe is made up of plasma. Plasma is the most abundant form of ordinary matter in the universe, and is often referred to as the fourth state of matter, the others being solid, liquid, and gas. Plasma consists of ions, electrons, and neutral bodies in both fundamental and excited states. Plasma generation can occur naturally and artificially via energy sources including electromagnetic fields, thermal, radiation, and bombardment by energetic species.

Plasma (from Ancient Greek πλάσμα) was first described in 1927 by American chemist Irving Langmuir as one of the four fundamental states of matter (i.e., solid, liquid, gas, and plasma) (Langmuir, 1928). Plasma was so named since the charged species that comprise the plasma can behave somewhat similarly to the corpuscular blood components that are bound by blood plasma (Morozov, 2012; Mott-Smith, 1971). Plasma is the most energetic and abundant state of matter, with as much as 99.9% of the universe's matter composed of plasma (Murawski, 2002). What makes plasma unique is its gaseous combination of ions, electrons, and neutral bodies in both fundamental and excited states. While plasma contains free charge carriers and is electrically conductive, commonly it is effectively electrically neutral from a macroscopic point of view. Unlike the three more familiar states of matter (gas, liquid, and solid), plasma is far less likely to be encountered at atmospheric conditions on the Earth and is typically artificially generated from a neutral gas. For example, plasma can be generated by the input of energy from electric fields, thermal sources, radiation, and beams (lasers, protons, UV photons, electrons) (Meichsner et al., 2012), as summarized in Figure 1.1.

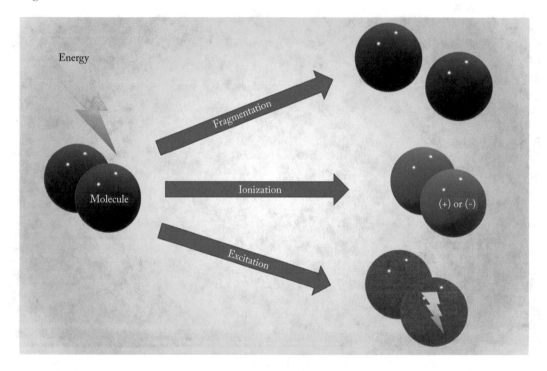

Figure 1.2: Processes of energy induced molecular fragmentation, ionization, and excitation.

As shown in Figure 1.2, atmospheric plasmas can be generated through the combination of energy-inducing fragmentation, ionization, and excitation of molecules. Such processes continue throughout the life of the plasma, leading to a wide variety of molecular and atomic species that can be highly reactive, electrically charged, energetically excited, or any combination of these states (Lu et al., 2014). Electric fields, energetic collisions, photons, or other mechanisms transmit energy to electrons, which in turn ionize some of the gas to generate a plasma of highly mobile electrons along with neutral, excited, and ionized gaseous species (Lu and Ostrikov, 2018; Ostrikov, 2005). Free electrons transmit energy to the neutral and ionized gaseous species by collisions. These collisions are probabilistic and can be divided into elastic collisions (where the internal energy remains constant for the gaseous species involved in the collision) and inelastic collisions (where energy is transmitted to the gaseous species and can excite the gas atom or molecule by changing the electronic structure of the gaseous species, or liberate an electron entirely resulting in ionization or additional levels of ionization) (Braithwaite, 2000). Most of the resulting excited gaseous species experience a very short lifetime (nano- to micro-second), returning to the ground state by emitting photons, thus exhibiting the characteristic plasma "glow." However, for metastable species, decay by emission is facilitated via energy transfer via collisions (Tendero et al., 2006). This results in metastable species in excited states having longer lifetimess. Generally, there are many possible mechanisms for the metastable decay of molecular ions, including vibrational predissociation of the parent ion in its ground electronic state, rotational predissociation, production of the parent ion in a metastable, and electronic excitation, etc. (Denifl et al., 2008).

During the second half of the 20th century, much plasma research has focused on space plasma physics and high-temperature thermonuclear fusion, yet significant efforts have developed low-temperature plasmas for terrestrial application (Graves, 1994). Other applications include magnetohydrodynamic (MHD) energy conversion, reentry boundary layers, arc jets, and plasma propulsion. Before glow discharge played a vital role in processing applications of electronic materials in the mid- or late-1970's, most interest involved gas discharge. These studies focused on various non-equilibrium plasma sources utilizing noble gases such as helium (He) and argon (Ar), as well as reactive gases such as O_2, H_2, CH_4, etc. (Blue and Stanko, 1969). Plasmas play an important role in many industrial sectors, especially for semiconductors (microelectronics devices and integrated circuits). These technologies include etching, sputtering, stripping, plasma-enhanced chemical vapor deposition, cleaning, plasma oxidation (anodization), plasma planarization, and plasma resist development (Graves, 1989; Shohet, 1991). Even today, research in non-equilibrium glow discharge plasma has yet to achieve equivalent levels to fields such as space electric propulsion, space plasmas, or plasma fusion science. One of the disadvantages of most plasma systems is the need to operate at low pressures which makes testing expensive and challenging. Atmospheric non-equilibrium plasma technologies, on the other hand, are commercially available and easy to operate and test. Pioneers in this new and important field include Kogelscaltz et al. of Sophia University (first proposed He atmospheric glow discharge), Roth et al. at the University of Tennessee (studies of atmospheric plasma glow discharges), Kogoma et al. at the University of Tokyo (atmospheric pressure glow discharge generation), Massines et al. at the University of Toulouse (the fundamental studies of N_+ and He plasmas), and Fridman et al. at Drexel University (innovative plasma medicine) (Fridman, 2008; Kanazawa et al., 1988; Pappas, 2011; Segui et al., 1996; Shenton and Stevens, 2001).

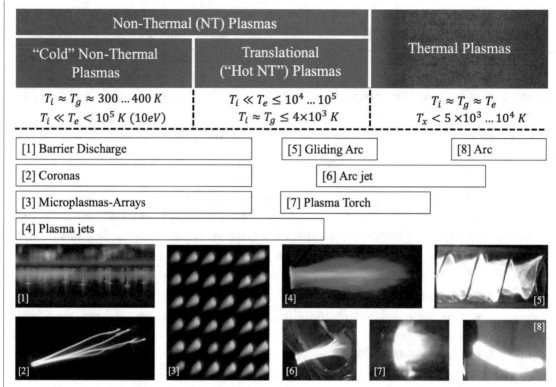

Figure 1.3: Atmospheric pressure and their classification within the thermal regime (Kalra et al., 2003; Park et al., 2014; Weltmann and Von Woedtke, 2011). Reproduced and modified from (Weltmann and Von Woedtke, 2011).

The properties of plasma (in terms of temperature or electronic density) change depending on the source and amount of energy supplied. According to Maxwell-Boltzmann thermodynamic equilibrium, plasma is divided into categories of *thermal* and *non-thermal* plasma. Figure 1.3 shows the classification of plasma generated at atmospheric pressure according to the thermal regime. Arc plasmas exhibit local thermodynamic equilibrium at the center with high temperatures for all species: ions, gas, and electrons (Griem, 1963). In contrast, a "cold plasma" is non-thermal (i.e., not in thermodynamic equilibrium) since the electron temperature is much higher than the temperature of ions and neutrals. Electrons in cold plasmas have typical mean temperatures of 10,000–100,000 K (~1–10 eV), while the temperatures of ions and neutral particles are of the order of 300 K (room temperature) to 1000 K (~ 0.03–0.1 eV) (Helmke et al., 2018). Due to this composition of cold ions and neutrals, cold plasma maintains high reactivity but imparts minimal thermal impact on the target surface, making cold plasma uniquely attractive for a wide range of applications. Characteristics of cold atmospheric plasma CAP include: (1) operation at atmospheric pressure; (2) mean electron temperatures sufficient for electron impact fragmentation, excitation, and ionization (>1 eV); (3) mean gas temperatures around room temperature; (4) heat energy transfer to the target below a deleterious level; and (5) current transfer to the target below significant induction of joule heating (Kong et al., 2009; Partecke et al., 2012; Van Gaens and Bogaerts, 2013).

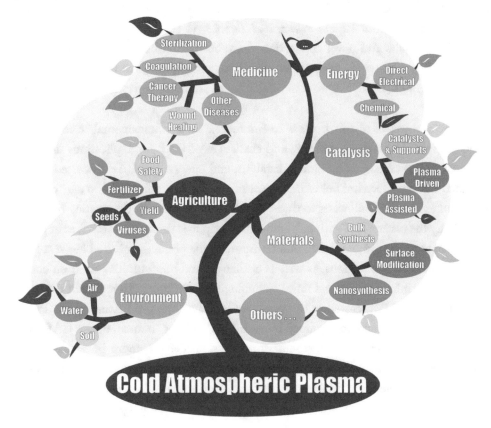

Figure 1.4: Cold Atmospheric Plasma Tree.

As an exceptionally interdisciplinary subject, CAP brings together many research fields, including atomic/molecular physics, statistical physics, electrodynamics, thermodynamics, heat transfer, fluid dynamics, material, electrical engineering, surface science, chemical engineering, chemistry, physics, and more recently, agriculture, foods, environment, energy, catalysis, biology, and medicine (Kortshagen et al., 2016; Lu et al., 2016; Neyts et al., 2015; Szili et al., 2018; Tendero et al., 2006; Winter et al., 2015). As discussed throughout this book, CAP contains reactive oxygen species (ROS), reactive nitrogen species (RNS), free radicals, ultraviolet (UV) photons, charged particles, electric fields, etc., which lead to a wide range of unique applications for various fields (Aerts et al., 2015; Attri et al., 2015; Bruggeman et al., 2017; Chen et al., 2021; Khlyustova et al., 2019a; Lu et al., 2016; Ono and Tokumitsu, 2015; Privat-Maldonado et al., 2019b; Takamatsu et al., 2014;

Vesel and Mozetic, 2017; Wende et al., 2018; Xian et al., 2013; Zarshenas et al., 2015; Zheng et al., 2021). CAP-related research is being published at an ever-increasing rate and is discussed prominently in The 2012 Plasma Roadmap (Samukawa et al., 2012). Prof. Peter Bruggeman, Prof. Uwe Kortshagen, and Prof. Mark J. Kushner organized more CAP experts to work on The 2017 Plasma Roadmap (low-temperature *Plasma Science and Technology*) (Adamovich et al., 2017). Recently, many important results and novel applications have been published in CAP subjects, including agriculture, medicine, materials, environment, catalysis, and energy (Bourke et al., 2018; Coutinho et al., 2019; Dou et al., 2018; Filipić et al., 2020; Lu et al., 2019; Neyts et al., 2015; Snoeckx and Bogaerts, 2017).

The main objective of this book is to summarize recent fundamental scientific aspects of CAP and the research progress, promise, and challenges of CAP for applications including agriculture, medicine, materials, environment, catalysis, and energy. The CAP discussed in this book are generated through electrical fields are typical of the order of a few mm^3 to cm^3 with electrical power ranging from a few milliwatts to watts. In this book, we present the existing body of knowledge and clarify the plasma applications in different disciplines through interactions of plasma with the matter. We consider the generation and transport of the plasma-generated key reactive species in the context of their applications, as well as challenges and opportunities for fresh concepts for expanded multidisciplinary study and novel technologies/applications. To represent the many fields and applications of CAP discussed in this book, we introduce the Cold Atmospheric Plasma Tree (Figure 1.4) with its applications grouped by each CAP field or "branch" of the CAP Tree. Each section of this book concludes with summary remarks, challenges, and a prospective outlook of CAP field discussed in that section. Additionally, a section of overall conclusions is provided at the end of this book. We intend for this summary of recent advances and applications of CAP provides a comprehensive understanding and guidance for colleagues, funding agencies, universities, industries, and government institutions.

CHAPTER 2

Cold Atmospheric Plasma (CAP)

2.1 CAP SOURCES

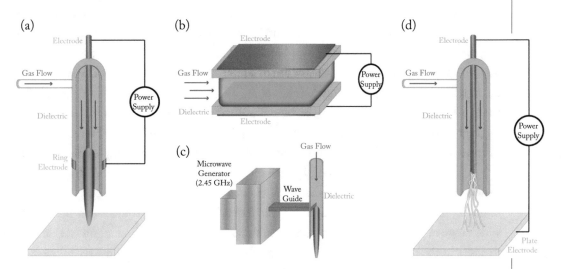

Figure 2.1: Typical cold plasma system configurations: (a) plasma jet, (b) dielectric barrier discharges (DBD), (c) microwave discharge, and (d) tip-plate discharges.

There are many categories to classify CAP sources, including plasma discharge mode, interaction characteristics with objects, and excitation frequency. CAP discharge modes can include any combination of plasma discharge such as glow discharge, dielectric barrier discharge (DBD), spark discharge, microwave discharge, and arc discharge. Discharges that operate at lower energy densities have applications for medicine, agriculture, foods, environment, materials, surface engineering, etc. (Bárdos and Baránková, 2010; Brandenburg, 2017; Fridman et al., 2005; Kim et al., 2015e; Laroussi et al., 2017; Lee et al., 2011; Nguyen et al., 2018; Ostrikov et al., 2013; Puač et al., 2018; Sakudo et al., 2020; Sanbhal et al., 2018; Setsuhara, 2016). Higher-energy discharges are generally employed for blood coagulation and incising tissues (Guild III et al., 2017; Nold et al., 2018). Classification according to electrical configuration and plasma/object interaction characteristics provides the general distinction, with various kinds of CAP developed for each. As shown in Figure 2.1, plasma jet, DBD, microwave discharge, and tip-plate discharge are common sources for cold plasma generating at atmospheric pressure. Plasma jets (Figure 2.1a) include different kinds of configurations that enable gas discharge in an open (non-sealed) electrode arrangement and project the discharge plasma reactive species (Nishime et al., 2017). These are often referred to as atmospheric pressure plasma jets (APPJs). DBDs (Figure 2.1b) are generated between two electrodes separated with the dielectric layers that reduce stray currents and spark formation (Moreau et al., 2008). Microwave-driven discharge plasma (Figure 2.1c) is generated by a magnetron and and employs a wave guide to transmit energy to the gas electrons. (Tolouie et al., 2018). Collisions that are primarily elastic occur between electrons and heavy plasma species and their interactions with the applied electric field result in heating of the electrons and slight heating of heavy species. Through these interactions, the electrons gain enough energy to produce inelastic exciting and ionizing collisions. Thus, the gas becomes partially ionized and becomes a plasma that supports electromagnetic wave propagation (Kabouzi et al., 2002). Applications of microwave plasma have driven the development of portable power modules: the integration of a microwave power amplifier chip with a power module to build a portable power source (Park et al., 2010). Tip-plate discharges (Figure 2.1d) use one sharp electrode (thin wires, edges, or tapred points) at atmospheric pressure where electric fields are high enough to accelerate electrons to energies sufficient to ionize the molecules and atoms of surrounding gas, leading to corona, spark, or arc discharges depending on operating conditions (Phan et al., 2017a). DBDs and APPJs are widely used to generate CAP, as the dielectric material between electrodes mediates the discharge to prevent the transition to a purely arc discharge and limit the discharge current through the electrodes. These plasma sources can be operated with a wide range of gases. For example, noble gases such as helium (He) and argon (Ar) are widely used to assist in the generation of CAP and typically exhibit a stable discharge with low gas temperatures. Other plasma sources convert just the ambient air (or surrounding gas) in close vicinity to the object surface into plasma (Kogelschatz, 2003). When the jet is exposed to air or air is mixed with assisting gases (sometimes called "feed" gases), gas-phase chemistry dominated by oxygen and nitrogen reactive species (ROS and RNS) is formed. These CAP reactive species are sometimes collectively identified as reactive oxygen and nitrogen species (RONS). ROS and RNS fluxes and composition are important factors for CAP applications and depend on many factors of the CAP device and operating conditions, including the power density, electrode spacing, electrode surface area, and gas composition and humidity.

Additionally, CAP sources can be classified with regards to excitation frequency due to its influence on the behavior of electrons and ions. Except for microwaves sources, two groups are proposed as follows: (1) direct current (DC) and low-frequency sources and (2) radio frequency (RF) wave sources (Tendero et al., 2006). DC and low-frequency sources can work in either pulsed or continuous modes. A pulsed power supply and the resulting plasma discharge is technically more complex than a continuous DC CAP source, potentially compromising the reproducibility of the process. However, a pulsed mode enables operation at high power levels but lower energies to reduce thermal stress on the device. For example, Fridman et al. developed a DBD for cold plasma treatment, assessing the safety and toxicity of the device for biomedical applications (Ayan et al.,

2008; Fridman et al., 2007). Although the DBD could be operated with pulsed or sinusoidal high voltages of 10–30 kV, the pulsed mode with pulse durations below 30–100 ns brought about better uniformity, and a no-streamer discharge with the possibility of non-damaging plasma treatment. For RF-driven sources, plasma can be generated at either high or low power, which influences the properties of the plasma and thus its potential applications. Types of devices with RF-driven sources include coupled plasma torch, CAP jet, cold plasma torch, micro-plasma, and so on (Park et al., 2012). Cheng et al. (2014) developed an AC-driven plasma jet operating at 30 kHz and voltage up to 10 kV. The discharge was generated with a 4.5-mm diameter quartz tube using a central 1-mm anode and outer wrapped cathode. Plume length was dependent on the voltage, frequency, and flow rate of the feeding gas, which was either He or a He/O_2 mix at 2-6 L/min; under these parameters, the discharge extended 5 cm from the quartz tube into the ambient atmosphere. Like many CAP sources, since the plume was at room temperature, it could be touched with bare hands without any thermal damage. Deng et al. (2007) presented a DBD, consisting of a dielectric tube wrapped with a metallic strip as the powered electrode, and a sample holder as the ground electrode placed 10 mm from the dielectric tub. The DBD was sustained by a peak voltage of 8 kV with a frequency of 30 kHz and employed He at a flow rate of 5 SLM alone or mixed with 25 SCCM of O_2.

2.2   CAP DIAGNOSTICS

By measuring various parameters, plasma diagnostics can elucidate important characteristics of discharge and the mechanisms of plasma-induced processes. The parameter space is very broad, ranging from the density of radicals to the velocity distribution of charged particles and even the population of electronically or rovibrationally excited states (Bruggeman and Brandenburg, 2013). Parameters' dynamic range can span many orders of magnitude, and temporal and spatial scales can significantly vary between configurations. In the last two decades, various diagnostic techniques have been developed to characterize CAP. Currently, these techniques include intensified charge-coupled device (ICCD) cameras, optical emission spectroscopy (OES), scattering techniques (such as Rayleigh-, Raman- and Thomson-scattering), laser-based methods (such as laser-induced fluorescence techniques), optical absorption spectroscopy measurements, mass spectrometry, gas chromatography, electron paramagnetic resonance spectroscopy, and some others (Adamovich et al., 2017; Chen et al., 2020d; Darny et al., 2017b; Hübner et al., 2015; Laroussi and Lu, 2005; Lu et al., 2016; Simoncelli et al., 2019; Van Doremaele et al., 2018; Vladimirov and Ostrikov, 2004).

ICCD cameras can be used to capture dynamic processes with picosecond time resolution at MHz acquisition rates, detecting/monitoring plasma discharge with pico- to nano-second precision. To gain such high time resolution, the measurement is performed by gating the acquisition time of the camera down to picoseconds and shifting the exposure with the electrical excitation phase. The obtained image is then an averaged emission image from a specific time point relative to plasma excitation (Schmidt-Bleker et al., 2015a). An example of such composite ICCD images for a typical CAP jet is shown in Figure 2.2a (I) alongside various time-dependent electrical and plasma parameters (Shashurin and Keidar, 2015). For each voltage period, the discharge consisted of a series of breakdown events that propagated the length of the He flow at 2×10^6 cm/s to a final decay after approximately 4 cm (Figure 2.2a (II)). The discharge current (I_d) also varied with this cycle, rising to 6–8 mA around the 1 μs time and decaying over 3–5 μs. Individual main discharge events remained stable for several μs after breakdown (Figure 2.2a (III), green bar).

Laser-based diagnostic techniques provide comprehensive information for plasmas (Darny et al., 2017a; Dilecce et al., 2015; Ghasemi et al., 2013). Laser-induced fluorescence (LIF) is based on the principle that wavelength-tunable laser radiation can shift molecules or atoms from one energy state to another, emitting fluorescent radiation during this transition. LIF can provide high spatial and temporal resolution information about the density of the ground and excited states of atoms and molecules. Schmidt et al. (2017b) investigated atomic oxygen in atmospheric pressure plasmas via LIF and compared femtosecond-two-photon-absorption laser-induced fluorescence (fs-TALIF) and nanosecond-two-photon-absorption laser-induced fluorescence (ns-TALIF). The performance of fs-TALIF had a number of differences when compared to ns-TALIF: (1) signal up to 15 times stronger; (2) fewer restrictions on pressure and operating conditions for measuring collisional quenching rates in situ; (3) lower power fluctuations (~2% of full scale vs. ~11%) and higher frequency repetitions rates (1–10 kHz vs. 5-50 Hz); (4) reduction in experimental uncertainty for atomic-oxygen densities (~±33% to ~±17%); (5) acquisition of 2D species distributions in minutes rather than hours; and (6) efficient and accurate measurements of atomic-species number densities. For plasma applications, the most dominant species as characterized by LIF are the hydroxyl radical (•OH) and the nitric oxide radical (NO•); others like oxygen, nitrogen, or water molecules are detected indirectly through quenching (Lu et al., 2016; Riès et al., 2014).

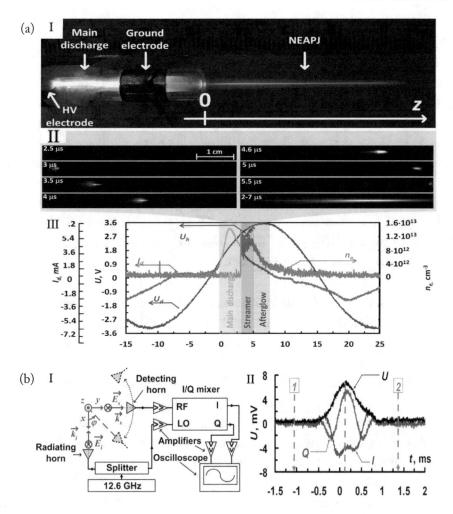

Figure 2.2: CAP diagnostics. (a) (I) Photograph showing a discharge assembly and CAP jet (25 kHz, 3.6 kV peak-peak, and He flow rate—11 L/min). (II) ICCD images were taken at 2.5–5.5 ms after initiation (100 ns exposure time) of streamer propagation for a CAP jet. The final image spans 2–7 μs for a 5 μs exposure. (III) Temporal evolution of discharge current, voltage, and average density in the streamer showing three stages of the streamer development: main discharge in the interelectrode gap (green bar), streamer growth (red bar), and the afterglow of the streamer channel (brown bar) are shown. Reproduced from (Shashurin and Keidar, 2015). (b) (I) Schematics of Rayleigh microwave scattering (RMS) experimental setup (upper) and (II) typical scattering signal measured by RMS setup for a calibrator bullet flying over the microwave horns along z-axis [in-phase (I) and quadrature (Q) components, and total amplitude of output signal U] (lower). A Teflon bullet of 8.5 mm length and 3.2 mm diameter was used (U_{HV}=3.8 kV for plasma jet and 15 L/min for He flow). Reproduced from (Shashurin et al., 2010).

OES is the most common diagnostic technique for the measurement of plasma sources (Bolouki et al., 2019). Because OES measures the radiation emitted by the relaxation of excited molecules or atoms in the plasma, it only characterizes those excited states. The information collected by OES can be derived from the spectral line widths, the absolute and relative line intensities, the emission continuum radiation, or the appearance of certain lines attributed to forbidden states (Coburn and Chen, 1980). The spectral width of an optical emission line provides valuables information about radiating species, and the dominant line broadening mechanisms are natural broadening, Doppler broadening, van der Waals broadening, Stark broadening, and resonance broadening (Ley, 2014; Pipa et al., 2015). Natural line broadening is homogeneous with its Lorentzian shape and, in the study of CAP, can be neglected due to its very small effect (on the order of 10^{-4}). Stark broadening can be applied to measure electron densities of CAP, though it is not reliable when electron density is less than 10^{13} cm^{-3}. Because of this, a collisional radiative model is an ideal method to determine the electron properties by relating the line ratios of selected emission lines (Nikiforov et al., 2015; Zhu and Pu, 2007).

Although fluorescence and emission measurements are generally non-invasive with high time and spatial resolution, the necessity of several assumptions regarding the model reduces their accuracy. Alternatively, scattering techniques that leverage elastic scattering from electrons (Thomson), heavy particles (Rayleigh), or inelastic scattering from molecules (Raman) are particularly useful. The Rayleigh microwave scattering (RMS) was first proposed by Shneider for measuring electron density in plasmas with structure smaller than microwave wavelength in 2005; it was then used to measure laser-induced avalanche ionization and resonance-enhanced multiphoton ionization in air and Ar plasma, respectively (Shneider and Miles, 2005; Zhang et al., 2007). Figure 2.1b shows the schematics of the RMS experimental setup and the experimental results. As stated by Shashurin et al., a homodyne method providing in-phase (I) and quadrature (Q) output was used via an I/Q mixer to detect the scattered signal (Shashurin et al., 2010). The resulting output signal U was calculated as $U = \sqrt{I^2 + Q^2}$. Using this technique to determine absolute electron density measurement over time revealed the presence of two consecutive breakdown events during a single voltage half-wave with the time delay between events inversely related to high-voltage amplitude. Plasma density in the streamer head was in the order of 10^{13} cm^{-3}. In addition, Gerling et al. characterized the electrical power distribution in the plasma jet through measuring the input power and voltage-current curves for the plasma in the effluent and the core plasma (Gerling et al., 2017). Other diagnostic techniques have also been used. As certain CAP applications may be incompatible with various methods of optical spectrometry and require other diagnostic methods, the identification of charged and neutral species such as those found in a plasma jet may be made by molecular beam mass spectrometry (Schmidt-Bleker et al., 2014). Separately, diffusion patterns for arbitrary flow fields can be calculated by mapping the pathways of these species, which can be applied to analyze absorption measurements, heat transport, and diffusion-based quenching of LIF signals

(Iseni et al., 2014; Reuter et al., 2012; Schmidt-Bleker et al., 2015b). Finally, electron paramagnetic resonance spectroscopy can be applied to detect radicals generated by plasma (Novák et al., 2013).

Although many mature diagnostic technologies for atmospheric plasma discharges have been developed, there are still many challenges (Adamovich et al., 2017). The stability and reproducibility of a plasma discharge have a huge impact on collecting high-quality data. The measurement precision is affected by not only the diagnostic equipment, but the plasma discharge itself. Current methods and technologies often lack the required sensitivity/selectivity or are not in situ performed with a sufficient temporal or spatial resolution. Further developments are certainly needed. The demand for identifying and measuring key parameters is growing, and the lifetimes of reactive species, transport behavior, and reaction schemes are of great interest (Bruggeman et al., 2017). In many applications, direct measurements of the parameters of interest are still not possible. Meanwhile, the plasma environments in application processes are extremely complex, and their reactions are still not fully understood. For example, when plasma interacts with liquid or tissue, it is challenging to resolve a very thin layer for observation. It also challenging to reveal the dynamics as species evaporating from the surface react with the incoming species created in the plasma flowing toward the surface. In addition, new applications of CAP will bring more challenges.

2.3   CAP SIMULATION AND MODELING

CAP can be described as partially ionized electrical discharges involving relatively frequent collisions among electrons and heavy particles (i.e. atoms, ions, and molecules). In CAP, most of the input power is transferred to electrons, resulting into a large thermodynamic discrepancy between heavy species and electrons (Lee et al., 2011; Trelles, 2018). CAP flows typically involve interactions among the working gas, any processing material, and the ambient gas environment, inducing large variation in the degree of ionization. Advances in experimental insights and improvements in algorithm accuracy and speed have led to an increase in the utilization of numerical modeling for CAP configurations. Different modeling approaches have been developed for CAP, such as fluid models, analytical models, non-equilibrium Boltzmann equations, Monte Carlo simulations, and particle-in-cell models (Capitelli et al., 2017; Cordes et al., 2016; Georghiou et al., 2005; Rat and Coudert, 2006; Yang et al., 2011; Zhang et al., 2015e). The advances and challenges faced by simulation of CAP can be broadly characterized in terms of numerical accuracy and model fidelity. Numerical accuracy involves the level of exactness of the solution of the model equations and thus characterizes the degree of resolution of temporal and spatial characteristics. Model fidelity concerns how comprehensive the model is in capturing underlying phenomena, with higher-fidelity models relying on fewer assumptions and containing a larger number of variables. Generally, accuracy is a major challenge for problems, with large variation in variable scale, whereas fidelity is a major challenge in a problem involving multiple physics disciplines. Further advancement in a field such as computational fluid dynamics (CFD) for plasma conditions would allow for the concurrent improvement of both accuracy and fidelity: CFD is already an effective tool for the design and analysis of plasma sources and processes in diverse applications in medicine, environment, agriculture, catalysis, material processing, and others. High/Low-order methods, which use high-fidelity models locally and low-fidelity models for background effects, and heterogeneous multiscale methods are potential approaches toward such advancements (Babaeva and Naidis, 2018; Chacon et al., 2017; Weinan et al., 2007). When implemented within CFD, these methods may help advance understanding plasma sources and broaden their applications.

As many of the desired outcomes depend on the interaction of reactive oxygen and nitrogen species with specimens, the mechanisms of plasma-generated ROS and RNS are critical to understanding. Table 2.1 summarizes reaction mechanisms of He plasma in air to show the many pathways toward ROS and RNS generation, which help scientists to understand the mechanisms of CAP in different applications.

Table 2.1: Reaction mechanisms of He plasma in air		
Reactions	**Rate coefficient**	**References**
$He_2^+ + N_2 \rightarrow 2He + N_2^+$ (B)	1.3×10^{-9} cm^{3}s^{-1}	(Ricard et al., 1999)
$e + He \rightarrow He + e$	*	(Kelly and Turner, 2014)
$e + He \rightarrow He^* + e$	*	(Kelly and Turner, 2014)
$e + He \rightarrow He^+ + 2e$	*	(Kelly and Turner, 2014)
$e + He^* \rightarrow He + e$	2.9×10^{-9} cm^{3}s^{-1}	(Stevefelt et al., 1982)
$e + He_2^+ \rightarrow He^* + He$	5.3×10^{-9}*Te$^{-0.5}$ cm^{3}s^{-1}	(Stalder et al., 2006)
$He^+ + 2He \rightarrow He + He_2^+$	8×10^{-32} cm^{6}s^{-1}	(Golubovskii et al., 2002; Ricard et al., 1999)
$He^* + 2He \rightarrow He + He_2^*$	2×10^{-34} cm^{6}s^{-1}	(Golubovskii et al., 2002)
$He^* + He^* \rightarrow e + He_2^+$	1.5×10^{-9} cm^{3}s^{-1}	(Golubovskii et al., 2002)
$He_2^* \rightarrow 2He + h\nu$	10^4s^{-1}	(Golubovskii et al., 2002)
$He_2^* + He_2^* \rightarrow e + He_2^+ + 2He$	1.5×10^{-9} cm^{3}s^{-1}	Golubovskii et al., 2002)
$e + O_2 \rightarrow 2e + O_2^+$	*	(Kelly and Turner, 2014)
$e + O_2 \rightarrow e + O + O$	*	(Kelly and Turner, 2014; Phelps, 1985)
$e + O_2 \rightarrow e + 2O^*$	*	(Kelly and Turner, 2014)
$e + O_2^+ \rightarrow 2O$	6×10^{-5}*Te^{-1} cm^{3}s^{-1}	(Kossyi et al., 1992)
$He^* + O_2 \rightarrow He + O_2^+ + e$	$2.54 \times 10^{-10}(T_g/300)^{0.5}$ cm^{3}s^{-1}	(Stafford and Kushner, 2004)
$He_2^* + O_2 \rightarrow 2He + O_2^+ + e$	$1 \times 10^{-10}(T_g/300)^{0.5}$ cm^{3}s^{-1}	(Cardoso et al., 2006)

$e + O_2 \rightarrow O + O^-$	*	(Kelly and Turner, 2014; Phelps, 1985)
$e + He + O_2 \rightarrow O_2^- + He$	$3.6 \times 10^{-31} Te[eV]^{-0.5}$ cm^{6}s^{-1}	(Murakami et al., 2012)
$O^- + O_2^+ \rightarrow O + O_2$	$3.464 \times 10^{-6} Tg^{-0.5}$ cm^{3}s^{-1}	(Kossyi et al., 1992)
$O_2^- + O_2^+ \rightarrow O + O + O_2$	1×10^{-7} cm^{3}s^{-1}	(Sakiyama et al., 2012)
$O^- + O_2^+ + He \rightarrow O + O_2 + He$	$3.12 \times 10^{-19} Tg^{-2.5}$ cm^{6}s^{-1}	(Kossyi et al., 1992)
$O_2^- + O_2^+ + He \rightarrow 2O_2 + He$	$3.12 \times 10^{-19} Tg^{-2.5}$ cm^{6}s^{-1}	(Kossyi et al., 1992)
$O^- + O_2 \rightarrow O_2^- + O$	1.5×10^{-18} cm^{3}s^{-1}	(Murakami et al., 2012)
$e + O_2 \rightarrow e + O_2^*$	*	(Kelly and Turner, 2014)
$O + O + X \rightarrow O_2 + X$	$2.15 \times 10^{-34} \exp(345/T_g)$ cm^{6}s^{-1}	(Peyrous et al., 1989)
$O + O_2 + X \rightarrow O_3 + X$	$3.4 \times 10^{-34} (300/T_g)^{1.2}$ cm^{6}s^{-1}	(Manion et al., 2015)
$O + O_2 + O \rightarrow O_3 + O$	$2.15 \times 10^{-34} \exp(345/T_g)$ cm^{6}s^{-1}	(Peyrous et al., 1989)
$O + O_2 + O_3 \rightarrow 2O_3$	$4.6 \times 10^{-35} \exp(1050/T_g)$ cm^{6}s^{-1}	(Peyrous et al., 1989)
$O + O_3 \rightarrow 2O_2$	$1.8 \times 10^{-11} \exp(-2300/T_g)$ cm^{3}s^{-1}	Peyrous et al., 1989
$O + O^* \rightarrow 2O$	8×10^{-12} cm^{3}s^{-1}	(Capitelli et al., 2013)
$O^* + X \rightarrow O + X$	1×10^{-13} cm^{3}s^{-1}	(Stafford and Kushner, 2004)
$O^* + O_3 \rightarrow 2O + O_2$	1.2×10^{-10} cm^{3}s^{-1}	(Capitelli et al., 2013)
$O^* + O_2^* \rightarrow O + O_2$	1×10^{-11} cm^{3}s^{-1}	(Kelly and Turner, 2014)
$O^* + O_2 \rightarrow O + O_2^*$	1×10^{-12} cm^{3}s^{-1}	(Capitelli et al., 2013)
$O_2^* + O_3 \rightarrow O + 2O_2$	$5.2 \times 10^{-11} \exp(-2840/T_g)$ cm^{3}s^{-1}	(Herron and Green, 2001)
$O_2^* + X \rightarrow O_2 + X$	2.01×10^{-20} cm^{3}s^{-1}	(Atkinson et al., 1997)
$O_3 + X \rightarrow O + O_2 + X$	$1.56 \times 10^{-9} \exp(-11490/T_g)$ cm^{3}s^{-1}	(Stafford and Kushner, 2004)
$O_3 + O_3 \rightarrow O + O_2 + O_3$	$1.65 \times 10^{-9} \exp(-11400/T_g)$ cm^{3}s^{-1}	(Peyrous et al., 1989)
$N_2^* + O_2 \rightarrow N_2 + O + O$	1.5×10^{-12} cm^{3}s^{-1}	(Popov, 2011)
$e + O_2 \rightarrow O^* + O + e$	*	(Phelps, 1985)
$O_2^- + O \rightarrow O_3 + e$	1.5×10^{-10} cm^{3}s^{-1}	(Sakiyama et al., 2012)
$O^- + O_2^* \rightarrow O_3 + e$	3×10^{-10} cm^{3}s^{-1}	(Sakiyama et al., 2012)
$e + N_2 \rightarrow N_2^{**} + e$	*	(Itikawa et al., 1986)
$e + N_2 \rightarrow N\_2^{\wedge(**)} + e$	*	(Itikawa et al., 1986)
$e + N_2 \rightarrow N + N + e$	*	(Itikawa et al., 1986)

$O_2^- + N_2^+ \rightarrow O_2 + N + N$	1×10^{-7} cm^{3}s^{-1}	(Sakiyama et al., 2012)
$N + O + X \rightarrow NO + X$	5.46×10^{-33}exp(155/T$_g$) cm^{6}s^{-1}	(Manion et al., 2015)
$N + O_2 \rightarrow NO + O$	4.4×10^{-12}(T$_g$/300)exp(-3270/T$_g$) cm^{3}s^{-1}	(Manion et al., 2015)
$N + O_3 \rightarrow NO + O_2$	5×10^{-16} cm^{3}s^{-1}	(Manion et al., 2015)
$N + OH \rightarrow NO + H$	4.7×10^{-11} cm^{3}s^{-1}	(Manion et al., 2015)
$N + O_2^* \rightarrow NO + O$	2×10^{-14}exp(-600/T$_g$) cm^{3}s^{-1}	(Van Gaens and Bogaerts, 2013)
$N + O_2\,(^1S) \rightarrow NO + O$	2.5×10^{-10} cm^{3}s^{-1}	(Uddi et al., 2009)
$N^* + O_3 \rightarrow NO + O_2$	1×10^{-10} cm^{3}s^{-1}	(Van Gaens and Bogaerts, 2013)
$N_2^* + O \rightarrow NO + N$	5×10^{-10} cm^{3}s^{-1}	(Shkurenkov et al., 2014)
$N_2^* + O \rightarrow NO + N^*$	1×10^{-12} cm^{3}s^{-1}	(Van Gaens and Bogaerts, 2013)
$N_2^* + O_3 \rightarrow NO + NO + O$	8.4×10^{-12} cm^{3}s^{-1}	(Van Gaens and Bogaerts, 2013)
$N_2^{**} + O \rightarrow NO + N$	5×10^{-12} cm^{3}s^{-1}	(Shkurenkov et al., 2014)
$N_2^{**} + O_3 \rightarrow NO + NO + O$	8.4×10^{-12} cm^{3}s^{-1}	(Van Gaens and Bogaerts, 2013)
$O_2^- + N \rightarrow NO_2 + e$	5×10^{-10} cm^{3}s^{-1}	(Sakiyama et al., 2012)
$NO + O + X \rightarrow NO_2 + X$	1×10^{-31}(300/T$_g$)$^{1.6}$ cm^{6}s^{-1}	(Manion et al., 2015)
$NO + O_2^* \rightarrow O + NO_2$	4.88×10^{-18} cm^{3}s^{-1}	(Manion et al., 2015)
$NO + O_3 \rightarrow O_2 + NO_2$	1.4×10^{-12}exp(-1310/T$_g$) cm^{3}s^{-1}	(Manion et al., 2015)
$NO+ + NO + O_2 \rightarrow NO_2 + NO_2$	2×10^{-38} cm^{6}s^{-1}	(Manion et al., 2015)
$NO_2 + O \rightarrow NO + O_2$	6.5×10^{-12}exp(120/T$_g$) cm^{3}s^{-1}	(Van Gaens and Bogaerts, 2013)
$NO_2 + H \rightarrow NO + OH$	1.28×10^{-10} cm^{3}s^{-1}	(Manion et al., 2015)
$NO_2 + O + X \rightarrow NO_3 + X$	9×10^{-32}(300/T$_g$)^2 cm^{6}s^{-1}	(Manion et al., 2015)
$NO_2 + O_3 \rightarrow O_2 + NO_3$	1.4×10^{-13}exp(-2470/T$_g$) cm^{3}s^{-1}	(Van Gaens and Bogaerts, 2013)
$NO + NO_3 \rightarrow NO_2 + NO_2$	1.6×10^{-11}exp(150/T$_g$) cm^{3}s^{-1}	(Manion et al., 2015)
$NO_3 + NO_3 \rightarrow NO_2 + NO_2 + O_2$	1.2×10^{-15} cm^{3}s^{-1}	(Person and Ham, 1988)
$NO + NO_2 + X \rightarrow N_2O_3 + X$	3.1×10^{-34} cm^{6}s^{-1}	(Manion et al., 2015)
$NO_2 + NO_2 + X \rightarrow N_2O_4 + X$	1.4×10^{-33} cm^{6}s^{-1}	(Manion et al., 2015)

$NO_2 + NO_3 + X \rightarrow NO_2O_5 + X$	3.6×10^{-30} cm^{6}s^{-1}	(Manion et al., 2015)
$He(2^3S) + He + N_2 \rightarrow 2He + N_2^+ (B) + e$	3×10^{-30} cm^{6}s^{-1}	(Ricard et al., 1999)
$He(2^3S) + 2He \rightarrow He_2(a) + He$	2×10^{-29} cm6s-1	(Ricard et al., 1999)
$O^* = O(^1D)$; $O_2^* = O_2(a1\Delta)$; $N_2^* = N_2(A^3\Sigma, B^3\Pi)$; $N_2^{**} = N_2(C^3\Pi$ and higher); $He^* = He(2^3P)$; $He_2^* = He_2(A^1\Sigma)$; X represents background gases (N_2, O_2, H_2, He, etc.); * = Rate coefficient to be obtained from Boltzmann's equation solution		

The interaction of CAP with liquids is of increasing importance in CAP applications, such as plasma-treated water (PTW), and has benefitted from insights from modeling and simulation. For example, Tian and Kushner (2014) reported a 2D computational investigation of plasma-generated ROS and RNS. The model employed in the paper was a 2D simulation, in which Poisson's equation, transport equations for charged and neutral species, the electron energy conservation equation, and radiation transport were solved by direct time integration. The reaction mechanisms for a He-fed atmospheric plasma at the water interface and in the water are summarized in Table 2.2. Electron transport and rate coefficients were calculated using Boltzmann's equation, and the Green's function propagator was employed for radiation transport, photon-induced ionization, and dissociation.

Table 2.2: Reaction mechanisms of He plasma at water interface and in water

Reactions	Rate coefficient	References
Water Interface		
$O + H_2O \rightarrow 2OH$	1.0×10^{-11}exp(-550/T$_g$) cm^{3}s^{-1}	(Peyrous et al., 1989)
$O + O + H_2O \rightarrow O_2 + H_2O$	2.15×10^{-34}exp(245/T$_g$) cm^{6}s^{-1}	(Peyrous et al., 1989)
$O + O_2 + H_2O \rightarrow O_3 + H_2O$	6.9×10^{-34}exp(300/T$_g$)$^{1.25}$ cm^{6}s^{-1}	(Peyrous et al., 1989)
$O_2^* + H_2O \rightarrow O_2 + H_2O$	3×10^{-234} cm^{3}s^{-1}	(Linstrom and Mallard, 2001)
$O_3 + H_2O \rightarrow O + O_2 + H_2O$	1.56×10^{-9}exp(-11490/T$_g$) cm^{3}s^{-1}	(Stafford and Kushner, 2004)
$N + O + H_2O \rightarrow NO + H_2O$	6.3×10^{-33}exp(140/T$_g$) cm^{6}s^{-1}	(Herron and Green, 2001)
$N + N + H_2O \rightarrow N_2 + H_2O$	8.3×10^{-34}exp(500/T$_g$) cm^{6}s^{-1}	(Capitelli et al., 2013)
$NO + O + H_2O \rightarrow NO_2 + H_2O$	1×10^{-31}exp(300/T$_g$)1.6 cm^{6}s^{-1}	(Herron and Green, 2001)
$NO_2 + NO_3 + H_2O \rightarrow N_2O_5 + H_2O$	2.8×10^{-30}exp(300/T$_g$)3.5 cm^{6}s^{-1}	(Atkinson et al., 1997)
$N_2O_5 + H_2O \rightarrow NO_2 + NO_3 + H_2O$	1×10^{-3}(300/Tg)3. 5exp(-11000/T$_g$) cm^{3}s^{-1}	(Atkinson et al., 1997)

$OH + OH + H_2O \rightarrow H_2O_2 + H_2O$	$6.9 \times 10^{-31}(T_g/300)^{-0.8}$ exp(-11000/T_g) cm^{6}s^{-1}	(Dorai and Kushner, 2003; Lukes et al., 2014)
$OH + NO_2 + H_2O \rightarrow HNO_2 + H_2O$	$7.4 \times 10^{-31}(300/T_g)2.4$ cm^{6}s^{-1}	(Herron and Green, 2001)
$OH + NO_2 + H_2O \rightarrow HNO_3 + H_2O$	2.2×10^{-30}exp(300/T_g)$^{2.9}$ cm^{6}s^{-1}	(Herron and Green, 2001)
$e + H_2O \leftrightarrow OH + H + e$	*	(Itikawa et al., 1986)
$e + H_2O \leftrightarrow OH + H^+ + e + e$	*	(Itikawa et al., 1986)
$e + H_2O^+ \leftrightarrow OH + H$	6.6×10^{-6}Te$^{-0.5}$ cm^{3}s^{-1}	(Rowe et al., 1988)
$O_2^- + H_2O^+ \rightarrow O_2 + OH + H$	1×10^{-7} cm^{3}s^{-1}	(Sakiyama et al., 2012)
$OH + OH + X \rightarrow H_2O_2 + X$	6.9×10^{-31}exp(300/T_g)$^{0.8}$ cm^{6}s^{-1}	(Manion et al., 2015)
$NO + HO_2 \rightarrow NO_2 + OH$	8.8×10^{-12} cm^{3}s^{-1}	(Manion et al., 2015)
$N_2O_3 + H_2O \rightarrow HNO_2 + HNO_2$	1.93×10^{-17} cm^{3}s^{-1}	(Manion et al., 2015)
$N_2O_4 + H_2O \rightarrow HNO_2 + HNO_3$	1.33×10^{-18} cm^{3}s^{-1}	(Manion et al., 2015)
$N_2O_5 + H_2O \rightarrow HNO_3 + HNO_3$	2×10^{-21} cm^{3}s^{-1}	(Manion et al., 2015)
$NO + NO_2 + H_2O \rightarrow HNO_2 + HNO_2$	5.55×10^{-34} cm6s-1	(Manion et al., 2015)
$NO + OH + X \rightarrow HNO_2 + X$	6×10^{-31} cm6s-1	(Manion et al., 2015)
$NO + HO_2 + X \rightarrow HNO_3 + X$	2×10^{-30} cm6s-1	(Manion et al., 2015)
$NO_2 + OH + X \rightarrow HNO_3 + X$	1.7×10^{-30} cm6s-1	(Manion et al., 2015)
$HO_2 + NO_2 \rightarrow HNO_2 + O_2$	1.2×10^{-13} cm^{3}s^{-1}	(Manion et al., 2015)
$NO_2 + OH + X \rightarrow ONOOH + X$	3.37×10^{-30} cm6s-1	(Manion et al., 2015)
$NO + HO_2 \rightarrow ONOOH$	6.09×10^{-11} cm^{3}s^{-1}	(Manion et al., 2015)
$He_2^+ + H_2O \rightarrow OH(A) + HeH^+ + He$	1.3×10^{-10} cm^{3}s^{-1}	(Ricard et al., 1999)
$He(2^3S) + He + H_2O \rightarrow 2He + H_2O^+ (X,A,B) + e$	3×10^{-34} cm^{3}s^{-1}	(Ricard et al., 1999)
$He_2(a) + H_2O \rightarrow 2He + H_2O^+ (X,A,B) + e$	6×10^{-10} cm^{3}s^{-1}	(Ricard et al., 1999)
$H_2O^+(A) + e \rightarrow OH(A) + H$	3×10^{-7} cm^{3}s^{-1}	(Ricard et al., 1999)
In-water		
$NO_{2(aq)} + NO_{2(aq)} + H_2O_{(l)} \rightarrow NO_2^- + NO_3^- + 2H^+$	1.0×10^8 M^{-2}s^{-1}	(Park and Lee, 1988; Seinfeld and Pandis, 2016)

$NO_{(aq)} + NO_{2(aq)} + H_2O_{(l)} \rightarrow 2NO_2^- + 2H^+$	2.0×10^8 M^{-2}s^{-1}	(Park and Lee, 1988; Seinfeld and Pandis, 2016)
$NO_2^- + H^+ \leftrightarrow HNO_2$	*	(Lukes et al., 2014)
$2HNO_2 \rightarrow NO + NO_2 + H_2O$	4.1×10^{-4} M^{-1}s^{-1}	(Park and Lee, 1988)
$HNO_2 + H^+ \rightarrow H_2NO_2^+ \rightarrow + NO^+ + H_2O$	5.0×10^3 s-1	(Lymar et al., 2002; Park and Lee, 1988)
$2NO_2 + H_2O \rightarrow NO_3^- + NO_2^- + 2H^+$	1.1×10^8 M^{-1}s^{-1}	(Seinfeld and Pandis, 2016)
$4NO + O_2 + 2H_2O \rightarrow 4NO_2^- + 4H^+$	10×10^9 M^{-2}s^{-1}	(Daiber and Ullrich, 2002)
$4NO_2 + O_2 + 2H_2O \rightarrow 4NO_3^- + 4H^+$	*	(Daiber and Ullrich, 2002)
$NO_2^- + H_2O_2 \rightarrow ONOO^- + H_2O$	1.1×10^3 M^{-2}s^{-1}	(Pryor and Squadrito, 1995)
$H_2O + NO \rightarrow ONOOH$	3.2×10^9 Ms^{-1}	(Goldstein and Czapski, 1995; Huie and Padmaja, 1993; Ischiropoulos et al., 1992)
$O_2^- + NO \rightarrow ONOO^-$	4.3×10^9 M^{-1}s^{-1}	(Katsumura, 1998)
$OH + NO_2 \rightarrow ONOO^- + H^+$	1.2×10^{10} M^{-1}s^{-1}	(Katsumura, 1998)
$2O_3 + H_2O + OH^- \rightarrow OH + HO_2 + 2O_2 + OH^-$	210 M^{-1}s^{-1}	(Staehelin and Hoigne, 1982)
$O_3 + H_2O_2 + OH^- \rightarrow OH + HO_2 + O_2 + OH^-$	5.5×10^6 M^{-1}s^{-1}	(Staehelin and Hoigne, 1982)
$NO_2^- + O_3 \rightarrow NO_3^- + O_2$	3.7×10^5 M^{-1}s^{-1}	(Hoigné et al., 1985)
$H_2O_2 + X \rightarrow OH + OH + X$	3.5×10^{-1} cm^{3}s^{-1}	(Manion et al., 2015)
$H_2O_2 + OH \rightarrow H_2O + HO_2$	2.7×10^7 M^{-1}s^{-1}	(Locke and Shih, 2011)
$H_2O_2 + O_3 \rightarrow O_2 + OH + HO_2$	0.065 M^{-1}s^{-1}	(Locke and Shih, 2011)
$H_2O_2 + H \rightarrow H_2O + OH$	3.6×10^7 M^{-1}s^{-1}	(Locke and Shih, 2011)
$H_2O_2 + e_{(aq)} \rightarrow OH + OH^-$	1.2×10^{10} M^{-1}s^{-1}	(Locke and Shih, 2011)
$H_2O_2 + NO_2^- \rightarrow H_2O + NO_3^-$	1.38×102^7 M^{-2}s^{-1}	(Anbar and Taube, 1954)
$H_2O_2 \leftrightarrow HO_2^- + H^+$	*	(Locke and Shih, 2011)
$O_3 + HO_2^- \rightarrow OH + O_2^- + O_2$	5.5×102^6 M^{-1}s^{-1}	(Staehelin and Hoigne, 1982)

Figure 2.3 shows the evolution of ROS densities in the gas gap (top) and 200 μm water layer (bottom). In general, reactive species generated within the plasma discharge are diffused onto the surface of the liquid, with solvency determined by atmospheric pressure as related by Henry's law constants. In some cases, it is specifically the processes within the gas phase that drive solvation. In Figure 2.3a, the OH in the gas phase was mainly produced by electron impact dissociation of water vapor and depleted by diffusion into water or formation of H_2O_2. H_2O_2 is more soluble at a given atmospheric pressure than OH and therefore more rapidly solvated into the water. The majority of the OH_{aq} was produced at the top of the liquid, which was also the entry point for OH diffusing from the gas phase and mutually reacted to form H_2O_{2aq} before reaching the underlying tissue in a 200 μm-thick water layer. Although H_2O_{2aq} was fairly stable in the absence of hydrocarbon in the employed model, H_2O_{2aq} could be photolyzed to OH by UV fluxes (250–300 nm) or could dissociate in the presence of ferrous ions, Fe^{2+} (France et al., 2007; Neyens and Baeyens, 2003).

In Figure 2.3b, O was dominantly generated in the gas phase during the current pulse by the dissociation of O_2, by direct electron impact, and by excitation transfer from N_2. The majority of O was consumed in the formation of O_3 in the afterglow and was regenerated by the next discharge pulse. The large density of O_{3aq} was consistent with the rapid rates of ozonation of water by remote plasmas. Modeling also indicated that H_2O_{2aq} and O_{3aq} were the dominant oxygen species reaching the underlying tissue. Another species, HO_{2aq}, is a transient species rather than a terminal species like O_{3aq} and was consumed by hydrolyzing into $H_3O^+_{aq}$ and O^-_{2aq}, or assisting NO_{aq} to form HNO_{3aq}. RNS fluences were dominated by NO^-_{3aq} and $ONOO^-_{aq}$. Even with such complex models as presented here, there is still much about the interaction of CAP-generated species ROS/RNS with various subjects.

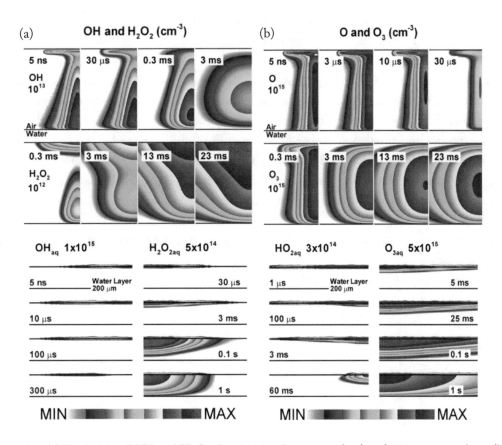

Figure 2.3: (a) Evolution of OH and H_2O_2 densities in the gas gap (top) and 200 μm water layer (bottom). OH and OH_{aq} densities are given for the first pulse and inter-pulse afterglow, while H_2O_2 and H_2O_{2aq} densities are shown accumulating after each pulse (0, 10, 20 ms) and during the terminal afterglow. (b) Evolution of ROS densities in the gas gap (top) and 200 μm water layer (bottom). O density is given for the first pulse and inter-pulse afterglow, while O_3 density is shown accumulating after each pulse (0, 10, 20 ms) and during the terminal afterglow. O_{3aq} and HO_{2aq} are shown in the liquid during the discharge pulses and through the terminal afterglow. All contours for (a) and (b) are plotted on a three-decade log-scale with the maximum density values noted for each set. Reproduced from (Tian and Kushner, 2014).

2.4   SUMMARY AND FUTURE DIRECTIONS

In summary, a wide range of CAP sources, diagnostics, and computational tools have been developed. Plasma dosage depends on plasma parameter and characteristics, and plasma diagnostics and modeling provide guidance to control plasma dosage for applications (Lu et al., 2016). There is much motivation to continue to use, expand, and improve these capabilities to support the wide range of applications and research challenges discussed in the forthcoming sections. For example, CAP sources are typically customized for their application, however, tunable CAP devices with well-established design and operation may be developed to accommodate wider ranges of applications (Adamovich et al., 2017). A higher understanding of commonality, at least for some subsets of applications, may reduce device cost, increase accessibility, and facilitate transparent collaboration and research. As the community continues to grow, concerted efforts toward establishing CAP diagnostics and simulation tools will also benefit overall community research efforts and provide needed tools for current and future applications (Laroussi et al., 2017). Over the last few years, many papers and important results have been proposed and published in each subject of this book, i.e., agriculture, medicine, environment, materials, catalysis, and energy (Bogaerts and Neyts, 2018; Bond, 1991; Bourke et al., 2018; Guo et al., 2020; Kaushik et al., 2019; Keidar, 2015). The following chapters of this book discuss each of these subjects and detail areas of particular interest for future research.

CHAPTER 3

Plasma Agriculture

The United Nations Department of Economic and Social Affairs projects that the global population will rise from some 7.7 billion in 2019 to 9.7 billion in 2050 (Gerland et al., 2014). With the world's rapid population growth, food demand is on the verge of increasing significantly. To meet this demand, global agriculture and the food industry will face more intense pressure than ever before. A growing concern in the agricultural sector is the threats posed by the use of pesticides, fumigants, and agrochemicals, which are harmful to the environment and health of most living creatures (Yadav and Devi, 2017). Furthermore, many targeted pests develop resistance to fumigants and pesticides, which leads to increases in their use by farmers in both developing and developed countries. Similarly, a number of reports have highlighted the problem of Fusarium fungus and related mycotoxins, which are major grain and fruit contaminants (Alshannaq and Yu, 2017; Richard, 2007). Thus, scientists must focus on technologies for controlling pest populations in fields as well as during post-harvest storage. At the same time, newly developed technologies should not leave any residual chemistry on food products and be able to modify the mode of action when necessary without incurring significant economic pressure. In short, efficient technologies are becoming increasingly necessary to reduce/arrest fungal growth, mycotoxin contamination, and pesticide residue in crops or harvested foods while reducing post-harvest losses, germination problems, and safety issues (Jermann et al., 2015; Knorr et al., 2011; Li and Farid, 2016; Misra et al., 2017). Moreover, it is important to develop sustainable food manufacturing and processing technologies that are environmentally friendly and consume fewer resources, including water and energy. Recently, CAP processing methods have been seen as having high potential because they work at low temperatures with short processing times without causing damage to crop, food, seeds, human beings, and the environment (Bourke et al., 2018; Cullen et al., 2018; Ito et al., 2018; Misra et al., 2016b; Moutiq et al., 2020; Puač et al., 2018). Plasma technologies' contribution in agriculture includes, but is not limited to, decontaminating seeds or crops intended for sowing or storing, disinfecting processing surfaces or tools, enhancing germination or growth, producing nitrogen-based fertilizers, remediation of soil, and reducing the invasion of pathogens (Charoux et al., 2020; Filimonova and Amirov, 2001; Mandal et al., 2018; Miao et al., 2020; Pina-Perez et al., 2020; Wang et al., 2020b).

3.1 SEEDS

Seeds require the highest probability of survival and the most efficient growth to increase productivity to meet the increasing demands of humans. Several studies have demonstrated a favorable effect of CAP treatment on seed germination and plant yields of a range of important crops, such as mung beans, maize, pepper, wheat, rice, oats, spinach, barley, cotton, lentil, rapeseed (Brassica napus L.), oilseed rape, green onion, tomato, soybean, thale cress, chickpeas, cucumber, and oilseed rape (Bormashenko et al., 2015; de Groot et al., 2018; Dubinov et al., 2000; Ji et al., 2016; Jiang et al., 2014b; Koga et al., 2015; Ling et al., 2015; Los et al., 2018; Puligundla et al., 2017; Sadhu et al., 2017; Shapira et al., 2017; Wang et al., 2017b; Waskow et al., 2018; Zahoranová et al., 2018). For example, Zhou et al. (2016) investigated the effects of atmospheric pressure N_2, He, air, and O_2 micro-plasmas on mung bean seed germination and seedling growth, as shown in Figure 3.1a. Seed germination percentage was strongly dependent on incubation time and the feed gas used, with the air and O_2 micro-plasma arrays demonstrating higher efficiency in enhancing seed germination. The authors also indicated that the acidic environment induced by plasma discharge in water may have promoted leathering of seed chaps, which enhanced the germination rate via stimulating the growth of hypocotyl and radicle. In a separate study, the interactions between plasma-generated ROS/RNS and seeds resulted in a significant acceleration of seed germination and an increase in the seedling length of mung beans (Li et al., 2005). Another factor leading to CAP-induced germination is the alteration of seed membrane properties to increase water uptake (Volin et al., 2000). CAP treatment, including the production of ROS and RNS, can be used to induce desirable changes in a broad spectrum of developmental and physiological processes in seeds, modifying seed coat structures, stimulating seed germination and seed growth, increasing the permeability of seed coats, and improving their resistance to stress and diseases (Jiafeng et al., 2014; Mitra et al., 2014; Selcuk et al., 2008; Šimončicová et al., 2018). For decontamination, Figure 3.1b shows a surface barrier discharge plasma experimental set-up for maize and the devitalization of surface pathogens as tested by Li et al. (Zahoranová et al., 2018). CAP treatment led to a significant reduction of pure cultures of toxicogenic *A. flavus*, phytopathogenic *A. alternata*, and seed-borne *F. culmorum* on maize seed surfaces. The most sensitive fungus was *F. culmorum*, where 60 s of CAP treatment was sufficient for total devitalization. Other scientists have demonstrated that CAP technology is a candidate for decontamination of foodborne pathogens documented as causative agents of outbreaks of foodborne illness associated with sprouts (Bormashenko et al., 2012; Butscher et al., 2016; Hertwig et al., 2017; Mošovská et al., 2018; Waskow et al., 2018; Ziuzina et al., 2013).

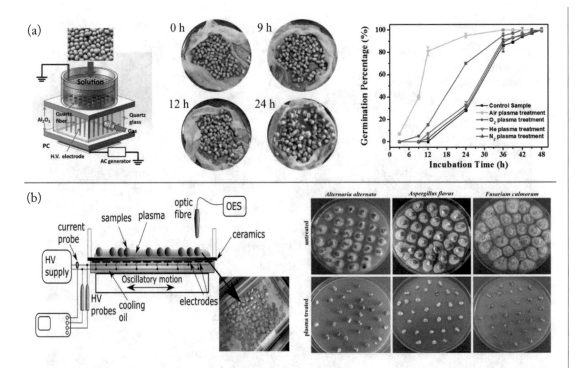

Figure 3.1: (a) Schematic diagram of an atmospheric-pressure micro-plasma array used to treat mung bean seeds in aqueous media (left); photographs of air-plasma-treated mung bean seeds at the incubation time of 0 h, 9 h, 12 h, and 24 h (middle); the germination percentage of mung bean seeds treated with He, N_2, air, or O_2 plasma as a function of incubation time (right). Reproduced from (Zhou et al., 2016). (b) Experimental set-up for the decontamination of maize via diffuse coplanar surface barrier discharge plasma (left); devitalization of various strains for intreated and treated maize samples (treatment duration in parens): *A. alternata* (300 s), *A. flavus* (300 s), and *F. culmoru*m (120 s). Reproduced from (Zahoranová et al., 2018).

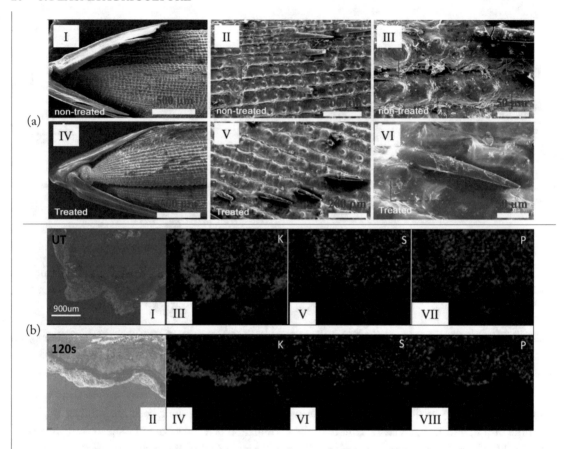

Figure 3.2: Modifications of the seed surface induced by cold plasma. (a) SEM micrographs of rice seeds, including the seed coat-lemma and palea-in uncreated (I-III) and CAP-treated (IV-VI) KDML105 rice. Blue dashed line windows highlight the microorganism living along the trench (II, III) and fragments of microorganisms (V, VI). Reproduced from (Khamsen et al., 2016). (b) Low-magnification SEM micrographs of nasturtium seed cuts before and after 120 s plasma exposure. Maps of the distribution of Potassium (K) (III, IV), Sulfur (S) (V, VI), and Phosphorus (P) (VII, VIII) before and after plasma treatments. Reproduced from (Molina et al., 2018).

Reactive oxygen and nitrogen species produced via CAP can decontaminate seeds or inactivate pathogenic fungi, stopping mycotoxin formation. Energized charged particles generated by CAP via electron-impact excitation can interact with the seed surface and lead to rupturing or etching the seed coat membrane (Mendis et al., 2000). In another study, CAP treatment induced the modification of microorganism distribution and surface morphology of a rice-seed leading to enhancement of the growth rate, however, there was no significant difference between the non-treated and treated samples when observed at low magnification, as shown in Figure 3.2a (I) and

(IV). Higher magnification revealed that, prior to treatment, the microorganisms were living along the surface trenches and a thin wax layer covered the seed surface (Figure 3.2a (II) and (III)). After CAP treatment, seeds had a cleaner surface, with the wax layer broken and only fragments of microorganisms and pathogens such as fungi remaining, indicating nanoscale modification of the surface morphology (Figure 3.2a (V) and (VI)). The wax layer is likely responsible for why the rice seed was naturally hydrophobic (water contact angle approximately 100°) with slow water absorption (30 min to complete the water uptake) prior to CAP treatment (Khamsen et al., 2016). There was no observable water contact angel after treatment (around 0°) with the water droplet instantaneously spreading over the surface, demonstrating one of the mechanisms by which CAP can mediate hydrophilic surface modification and enhance water imbition. In addition, CAP-generated RNS including nitrate radical and nitric oxide may proliferate on the seed surface to improve nitrogen fixation and provide an important macronutrient to the seed (Raizer and Mokrov, 2013). The EDX mapping of elements on slices of plasma-treated nasturtium seeds extending from the surface to several microns inside the pericarp is shown in Figure 3.2b. In the outer layer to a depth of approximately 300 μm, element potassium (K) was enriched, while other elements were either unaffected or had their concentrations slightly depleted. Plasma treatment of quinoa and pepper seeds in prior research has shown similar migration of K toward the surface (Gómez-Ramírez et al., 2017; Štěpánová et al., 2018). Using a technique known as Attenuated Total Reflection for Fourier Transform Infrared Spectroscopy to analyze treated nasturtium seeds, Molina et al. also report that the resultant normalized spectra contained peaks and bands associated with the stretching, bending, or other deformation of various molecular bonds: CH-caused deformations and bending vibrations of OH (1473–1296 cm^{-1}); C-O-caused stretching of carbohydrates (cellulose and hemicellulose) (1246 cm^{-1}), CH (2990–2744 cm^{-1}), or C = O of carbonyl groups in lipids (1732 cm^{-1}), aromatic stretching band of lignin (1602 cm^{-1}); and C-O-caused deformation of carbohydrates and lignin (1187–857 cm^{-1}) (Lisperguer et al., 2009; Molina et al., 2018; Yu et al., 2003). Extending the CAP treatment time did not change the shape of each peak or band beyond decreased stretching vibration in OH groups, nor did CAP treatment include any significant chemical reactions on the surface (Molina et al., 2018). In summary, cold plasma has been shown to improve various physicochemical and physiological properties of different seeds, affecting plant germination and growth (Bourke et al., 2018). The physicochemical properties include moisture content, protein concentration, chlorophyll levels, hydrophobicity, enzymatic activity, nitrogen prevalence and availability, soluble phenol content, and wettability, while physiological properties include fresh weight, vigor, germination rate, overall yield, and growth rate (Mitra et al., 2014). To enhance seed growth, the plasma treatment dosage and chemical composition as determined by power input, working gas, and treatment time, should be adequately evaluated separately for each type of seed being treated.

3.2   CROPS

Over the last several decades, CAP has been applied to crops due to its ability to deliver myriad reactive species to living organisms with minimal accompanying damage (Koga et al., 2015; Puač et al., 2018; Selcuk et al., 2008). CAP can enhance the growth of crops as shown in Figure 3.3. Growth enhancements of up to 40% were observed in sprouts receiving plasma-treated water (PTW) after a retention time of 30 min (Figure 3.3a I) (Ito et al., 2018). For direct treatment, Sarinont et al. investigated the effects of plasma irradiation using various feeding gases (air, O_2, NO (10%) + N_2, N_2, He, and Ar) on the growth of *Raphanus sativus L* (Sarinont et al., 2016). Among all treatment, air plasma irradiation was the most effective for growth enhancement, with an increase of approximately 42% (as shown in Figure 3.3a II). The combined effects of plasma treatment of seeds and irrigation with PTW for pepper crops long-term growth were studied by Sivachan-diranab and Khacef (Sivachandiran and Khacef, 2017). CAP showed a positive effect on the crops: the most promising results were obtained for 10-min seed treatment and watering with deionized water activated by 30-min plasma treatment (as shown in Figure 3.3a III). Leaves of treated pepper plants appeared healthier than the controlled samples from visual analysis.

In general, PTW is becoming an increasingly important topic in the field of plasma science/technology, and its applications include agriculture, medicine, and environment (Bang et al., 2020; Jablonowski and von Woedtke, 2015; Kaushik et al., 2018; Khlyustova et al., 2019a; Liu et al., 2020a; Lukes et al., 2014). While more details about the generation of PTW covered elsewhere in this book, we discuss PTW for crops in this section and other applications in the following chapters. Free radicals and reactive species generated in the atmosphere and at the plasma-water interface can penetrate side interface and enter so-called "bulk water" to react or recombine, giving rise to the generation of more stable reactive species (as shown in Figure 3.3b and Table 2.1) (Brug-geman et al., 2016; Van Gils et al., 2013; Vanraes and Bogaerts, 2018; Wende et al., 2018). PTW can thus not only inactivate a variety of microorganisms but also function as a fertilizer (Judée et al., 2018; Kamgang-Youbi et al., 2018). For organism inactivation, it has been established that PTW's bactericidal activity results from the combined action of a low pH and a high positive oxidation-re-duction potential (Oehmigen et al., 2010). With as short as 200 s of plasma treatment, PTW's pH decreased to 5 and H_2O_2 concentration increases to 2 mg/L, noticeably affecting seed germination and crop growth (Bruggeman and Leys, 2009; Shoemaker and Carlson, 1990). For example, an essential nutrient for crops is nitrogen as present in plasma-generated NO_3- as it is a constituent of proteins, chlorophyll, amino acids, and other metabolites/cellular components. In addition, plant growth can be improved through the activation of genes and proteins by an appropriate dose of H_2O_2 (Sivachandiran and Khacef, 2017).

Figure 3.3: Cold plasma effect on crops and its reactions at interface. (a) Growth enhancement of sprouts cultivated in untreated/plasma-treated water (I), Reproduced from (Ito et al., 2018); photograph of air-plasma irradiated Radish sprouts after 7 days' cultivation by air plasma irradiation (II), Reproduced from (Sarinont et al., 2016); and long-term effect of plasma-treated water and plasma seed treatment on pepper crops (PTW-30: deionized water was activated by plasma 30 min; P10: Petri dish contained 10 seeds; TW: tap water) (III), Reproduced from (Sivachandiran and Khacef, 2017). (b) Photograph of He plasma interaction with water and its mechanisms.

Oxygen is another element critical to plant growth. Crops engage in active oxygen scavenging through mechanisms that can either be enzymatic (including glutathione peroxidase, peroxidase, ascorbate peroxidase, and superoxide dismutase) or non-enzymatic (including carotenoids, flavonoids, and ascorbic acid) (Apel and Hirt, 2004). Direct CAP irradiation and PTW (indirect) treatment enhance the enzymatic activity of ascorbate oxidase, peroxidase, and ascorbate peroxidase while reducing zeatin and abscisic acid (Li and Xue, 2010). The oxygen-scavenging abilities of plants are increased by the delivery of sufficient plasma-generated ROS to seeds and crops, increasing seed germination rate, and promoting the growth of crops. Overall, PTW is an environmentally friendly disinfectant and fertilizer, when compared to more traditional chemical sanitizers or fertilizers.

3.3   FOODS

There has been an increasing market demand for safe, high-quality, and minimally processed food products worldwide. The risk associated with the transmission of foodborne pathogenetic infections has become a global issue of concern to the entire food industry. In 2011, the European Union alone reported over 5,600 foodborne outbreaks leading to almost 70,000 human cases, with over 7,000 hospitalizations and 93 deaths. The prevalence of health-associated outbreaks, often linked to the consumption of food products, implies that current sanitation procedures have limited efficacy against bacterial pathogens and creates a demand for more efficient decontamination techniques (Niemira, 2014; Olaimat and Holley, 2012; Ölmez and Temur, 2010; Warning and Datta, 2013). CAP treatment as an alternative, ecologically friendly process for the preservation of foods meets the increasing market demand for minimal processing while enabling safe and high-quality food products with low cost (Pinela and Ferreira, 2017; Rana et al., 2020; Sun, 2014). The usefulness of CAP processing for microbial inactivation while maintaining the quality of fresh products has been known for years (Aboubakr et al., 2020; Mahnot et al., 2020; Song and Fan, 2020; Ziuzina et al., 2020). General application of CAP for microbial destruction on different food substrates like fruits, meat products, cheese, vegetables, etc., have previously been investigated with positive results (Kim et al., 2019; Misra and Jo, 2017; Niemira, 2012; Perni et al., 2008; Song et al., 2009). It has been applied to inactivate the endogenous enzymes responsible for browning reactions, particularly *poly-phenol oxidase* and *peroxidases* (Pankaj et al., 2013; Surowsky et al., 2013). A review by Coutinho et al. (2018) attributed the antimicrobial effects of CAP to ROS/RNS interactions. These interactions result in "strong oxidative effects on double bonds in the lipid bi-layer of the microbial cell", and, additionally, disrupt macromolecule transport for the cell both internally and externally. In addition, several other mechanisms for antimicrobial effects also exist: other atoms or ions, emissions (UV), electrons, metastable, and electronically/vibrationally excited molecules.

As a food-processing aid, the potential of CAP has been demonstrated for a range of processes and products from the seed stage to the post-harvest stage and from market to disposal, as shown in Figure 3.4. The potential applications are extensive, including: (1) sterilization of seeds while in storage; (2) enhancing seed germination and yield; (3) generating RNS in water for use as a fertilizer; (4) adding ROS to soils and lowering their pH to reduce pathogen invasion; (5) removal of volatile organic compounds in greenhouse facilities and air cleaning/sterilization; (6) sterilizing and puri-fying water for post-harvest sanitation; (7) disinfection of products prior to packaging; (8) cleaning and sterilizing the packaged product storage facility and transportation vehicles; (9) control of pests and pathogens during retail storage and display; (10) removal of ethylene from air to reduce the rate of aging/spoiling; (11) sterilization of cutting boards, knives, and other food processing equipment both at home and in food processing facilities or grocery stores; (12) plasma-assisted destruction of hazardous waste and/or waste-to-energy conversion of the non-hazardous food wastes; and others

(Cullen et al., 2018; Ekezie et al., 2017; Guo et al., 2015; Hertwig et al., 2018; Park et al., 2013; Pérez-Andrés et al., 2020; Pignata et al., 2017). In food processing, food packaging plays one of the most important roles by maintaining many of the benefits of processing until the time of consumption (Marsh and Bugusu, 2007; Zhao et al., 2020). An interesting approach was demonstrated by Klockow and Keener, where prepackaged whole spinach leaves inoculated with *E. coli O157:H7* were exposed for 5 min to cold plasma and subsequently stored for 0.5–24 h (Klockow and Keener, 2009). Post-treatment storage for 24 h resulted in the largest reductions in the bacterial presence (3–5 log CFU/leaf) and was determined to be a critical treatment parameter for maximizing the efficiency of bacterial inactivation when using cold plasma. Clarke et al. investigated the use of packing materials made from antimicrobial gelatin-coated low-density polyethylene after surface activation via cold plasma treatment, which was shown to be effective against bacterial species evaluated on beef steaks (Clarke et al., 2017). CAP has also been considered for extension of food shelf-life, with preliminary results demonstrating that it improves food shelf-life and is an ideal technology for ready-to-eat foods such as fresh meat, fruits, and vegetables (Moon et al., 2016; Patange et al., 2017; Tappi et al., 2016). Meanwhile, cold plasma has been applied for insect control in-store products, increasing larval/pupal mortality and reducing the adult emergence rate due to stress caused by the action of ROS generated during the treatment (Abd El-Aziz et al., 2014; Mohammadi et al., 2015).

CAP has been successfully employed for degradation of mycotoxins (via the high oxidizing potential of ROS) which can contaminate seed, grains, fruits, or crops and pose high risks to human health(Hojnik et al., 2017; Misra et al., 2019a). This degradation has been demonstrated for many differe nt organophosphorus and organochlorine pesticide compounds and on various substrates. From the perspective of human consumption, plasma-treated dried food products have been found to have no difference in sensory acceptability, regarding appearance, color, flavor, taste, and texture (Choi et al., 2017; Gurtler and Keller, 2019). Other studies have reported plasma-treated meat or fruits to be lighter in color or slightly green, possibly due to the bleaching effect of ROS (Fröhling et al., 2012; Ma et al., 2015). CAP can also modify the functionality of wheat flour toward generating a stronger dough and accelerating lipid peroxidation and reduce the cooking time of brown rice through increasing grain hydrophilicity and modifying rheological properties (Bahrami et al., 2016; Thirumdas et al., 2017; Yodpitak et al., 2019). CAP maintains food quality and consumer acceptability during storage due to the inactivation via plasma exposure of enzymes involved in the browning of fruit and vegetables. Plasma exposure initiates changes across a range of food chemicals, including the modification of amino acid residues in proteins, the oxidation of sugars to organic acids, the inactivation of enzymes, and the peroxidation of lipids in general and unsaturated fatty acids in particular (Han et al., 2019; Okazaki et al., 2014; Park et al., 2015; Zhang et al., 2015b). Han et al. conducted experiments to evaluate the acute and subacute toxicity of plasma-treated food and indicated that consumption of the edible film treated with plasma did not induce any significant toxicological effects (Han et al., 2016b). Food allergies affect approximately 10% of people around the world, and the top 8 food sources triggering allergic reactions are tree-nuts, peanuts, eggs, milk, fish, crustacean/shellfish, wheat, and soy (Bourke et al., 2018). The best prevention method is the avoidance of the food allergen with individually variable threshold doses. Cold plasma shows potential as a tool to reduce the immunoreactivity of food allergens in foodstuffs and processing environments, which is particularly suitable for those allergens that are recalcitrant to standard processing due to their thermostability (Meinlschmidt et al., 2016; Segat et al., 2016). Overall, research has shown that CAP is a potent tool for increasing the efficiency of growth and processing for different stages of food production, with system design capable of being optimized for enhancement of various microbiological, physiological, and chemical quality characteristics of different types of foods.

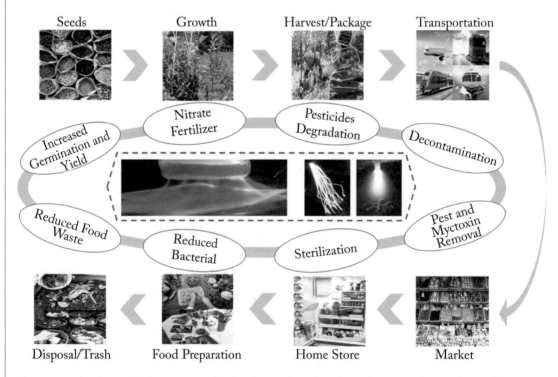

Figure 3.4: Applying CAP to improve food safety and quality through increased growth and decontamination of food at seed, growth, harvest, and post-harvest stages. The results are increased food yield, efficient degradation of pesticides, reduced food waste, and degraded mycotoxins as well as inactivated pests.

3.4   SUMMARY AND FUTURE DIRECTIONS

In summary, although the potential for CAP as a tool in agriculture is high, there are still gaps in the research that need to be addressed before CAP processes can be efficiently utilized in large-scale applications (Brandenburg et al., 2019). Replacement of existing sterilization, treatment, and other procedures in the industry will require demonstrated improvements in efficiency and reliability, for which specific knowledge of the chemical reactions taking place will be needed (Chen et al., 2020e; Misra et al., 2016b). For example, the underlying mechanisms by which seeds benefit from CAP need to be identified in detail, i.e., while mechanisms have been generally proposed, no isolation of the specific effects of individual reactions has been undertaken, partly due to the complexity of plasma discharge. Particular effects, such as physical changes to seed coatings, have been observed directly and well-documented but still lack underlying chemical processes that can be tuned for improved efficiency (Ekezie et al., 2017). While limited negative effects from plasma treatment have been reported, few industrial-scale experiments tackling the full lifecycle from seed through germination and growth to harvest have been performed. As a result, there may be unforeseen consequences or effects on taste, nutritional content, or usability for produce grown using plasma-assisted methods. As one example, CAP has been used to successfully inactivate various enzymes and change protein structures in a targeted manner; detailed understanding and careful study are needed to ensure that similar effects do not occur unintentionally during other CAP treatments (Han et al., 2019; Misra et al., 2016a). To isolate the varied mechanisms of plasma treatment and develop industrial-scale methods of application, the interdisciplinary collaboration will almost certainly be required. The roles of physicists and agricultural scientists are immediately obvious, but additional expertise in other fields will need to be brought into play: materials science for a detailed understanding of hydrophobicity mechanisms and seed structures; biochemists to isolate and identify specific enzymic or structural changes within treated organisms; data scientists to isolate specific effects from lists of complex interactions; engineers to design and build processing systems; and so on. While such a list of participants may seem daunting, it is the complex nature of CAP interactions that drives such profound and far-reaching results.

CHAPTER 4

Plasma Medicine

Plasma medicine is an innovative and emerging interdisciplinary research field that attracts researchers from different fields such as engineering, chemistry, physics, biology, and medicine (Babaeva and Naidis, 2018; Fridman et al., 2008; Graves, 2014a; Laroussi, 2018, 2019; Park et al., 2012; Tanaka et al., 2017; Von Woedtke et al., 2013). It brings technological challenges for developing plasma-based medical devices and raises fundamental questions regarding the mechanisms of interaction between plasma and living tissue (Judée and Dufour, 2019; Stancampiano et al., 2019; Stoffels et al., 2008). Originally, plasma attracted considerable attention in the medical field because of its promising heat effect for such uses as electrosurgery, blood coagulation, and sterilization of heat-sensitive medical instruments (Iza et al., 2008; Raiser and Zenker, 2006). Recently, the co-existence of hot electrons, cold neutral gas, and ions has drawn scientific interest toward CAP and its non-thermal effects (Bárdos and Baránková, 2010; Ten Bosch et al., 2019). Although plasma has been known for decades to have biocidal effects, the successful killing of bacteria with plasma was only reported in 1996 (Laroussi, 1996). Preliminary experimental and clinical data suggest that CAP allows efficient, contact-free, and painless disinfection without damaging healthy tissue, even in microscopic openings. New horizons have been opened for sterilization, coagulation, wound healing, the treatment of skin infections, tissue regeneration, dental treatment, cancers, and many more diseases (Chen et al., 2016d; Huang et al., 2020; Mateu-Sanz et al., 2020; Metelmann et al., 2018b; von Woedtke et al., 2019). The key roles of ROS and RNS in plasma medicine have become clearer in the past several years. ROS and RNS are biologically and therapeutically active agents, which are also well known to play important roles in a wide variety of intercellular and intracellular processes (Bauer and Graves, 2016; Graves, 2012, 2014b, c; Haralambiev et al., 2020; Keidar, 2018; Privat-Maldonado et al., 2019b; Szili et al., 2018; von Woedtke et al., 2019). CAP affects cells through the programmable process of apoptosis, which is a multi-step process inherent to cells in the human body that results in cell death (Bekeschus et al., 2020b; Evan and Vousden, 2001; Kerr et al., 1972). The recent demonstrations of CAP in the treatment of living cells and tissues have been very promising and reveal the potential for novel applications of CAP. We will discuss more recent advances in applications of CAP in medicine in detail.

4.1 DISINFECTION

The development of energy-efficient and effective approaches for the damage of biologically dangerous contaminants at room temperature in liquids, gases, and on the surfaces of bodies is a big challenge for modern science. In addition to disinfection, the major application is the protection of various industrial materials, electronic devices, and types of equipment against biodegradability and biocorrosion. Disinfection of objects comprises of damage or removal of microorganisms, including spores, vegetative cells, viruses, bacteria, etc. Traditional disinfection methods utilize strong chemicals, heating in dry/humid environments, radiation, and filtration, which are time-/labor-consuming and expensive (Akishev et al., 2008). CAP has unique characteristics containing numerous biochemically active agents, which can be created in liquid or gas at ambient temperature (Bourke et al., 2017; Chen et al., 2016d). The initial application of CAP in the field of biology was decontamination, and the inactivation effects of plasma on various fungi, bacteria, and viruses were reported (Laroussi, 1996).

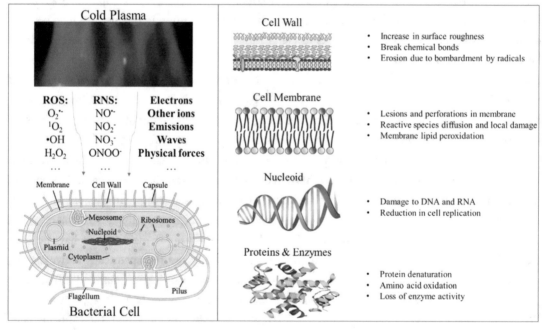

Figure 4.1: Overview of CAP mechanisms involving in plasma action on bacterial cell structures and leading to loss of functionality and sterilization. During treatment, bacterial cells are exposed to continuous bombardment with reactive components generated by plasma, inducing damage on cell walls, cell membranes, nucleoids, proteins, enzymes, etc.

Four basic mechanisms are driven by plasma treatment and contribute to bacterial cell death, namely etching cell walls, damaging cell membranes, destruction of genetic material, and protein/enzyme denaturation. Figure 4.1 illustrates the mechanisms of CAP-caused microbial inactivation. Ziuzina et al. employed electron microscopy to image the change in morphology of bacterial cells or their lysis under cold plasma treatment (Ziuzina et al., 2014). The resulting plasma-induced bacterial cell leakage and cell functionality loss was the compounded effect of generated reactive species, which vary with factors including the plasma source, feeding gas types, operating conditions, the nature of the product/substrate, and the micro-organism itself (Bartis et al., 2013; Ehlbeck et al., 2010; Misra and Jo, 2017). Yusupov et al. (2013; 2012) used computer simulations to analyze plasma-induced damage to bacterial cell walls. Their study highlighted the breakage of structurally important bonds, such as C-C, C-O, and C-N. Further, intracellular pH may have been lowered by generated ROS and RNS diffusing into the bacterial cell, inactivating it due to the absence of pH homeostasis (Laroussi and Leipold, 2004). Damage to or disruption of the bacterial cell membrane has also been studied. For example, mass transport through the membrane becomes dysregulated by oxidative damage to membrane lipids (Padan and Schuldiner, 1987). Separately, charged particle accumulating on the microorganism's surface also damages the cell membrane through electrostatic disruption (Hertwig et al., 2017). The electrostatic forces from such an accumulation can exceed the tensile strength of the membrane, leading to its rupture (Mendis et al., 2000). ROS also play a part in membrane disruption, as they can drive protein denaturation and cell leakage (Misra et al., 2016a). Such reactive species can induce oxidation of amino acids, nucleic acids, and unsaturated fatty acid peroxides through interaction with membrane lipids, leading to changes in the membranes' function (Critzer et al., 2007). Misra et al. reported that nucleic acids oxidized to 8-hydroxy-2 deoxyguanosine, and amino acids oxidized to 2-oxo-histidine. Changes in the membrane directly affect the nucleoid primarily via pore formation and breakdown of the interactions between nucleoid and membrane proteins (Moreau et al., 2008). The direct effects of ROS and RNS on the nucleoid are to inhibit cell replication by forming strand breaks and thymine dimers.

Other CAP-generated emissions, such as UV, are also responsible for cellular damage. For example, UV emission can, similarly to ROS and RNS, induce the dimerization of thymine bases in bacterial DNA strands with similar effects on the ability of bacteria to replicate (Laroussi, 2002). UV emission also can damage the DNA of bacterial spores, with increased UV irradiation corresponding to increasing damage to spore DNA (Kim et al., 2015d; Setlow, 2007). Moreover, UV emission is able to induce erosion of the cell, atom by atom, due to the breakage of chemical bonds in microorganisms, which is then followed by the formation of volatile by-products such as CH_x and CO from unbound atoms (Moisan et al., 2001; Schlüter and Fröhling, 2014). The volatile compounds that form when microorganisms are exposed to intense plasma radicals provoke lesions on the surface that the cell cannot repair and result in cell death.

Moreover, disinfection by CAP can be an alternative to conventional and/or traditional disinfection methods for coronavirus disease 2019 (COVID-19) (Chen and Wirz, 2020). COVID, an infectious disease caused by severe acute respiratory syndrome coronavirus 2 (SARS-CoV-2), has induced a once-in-a-century pandemic, seriously threatening millions, if not billions, of people (Guan et al., 2020; Wu and McGoogan, 2020). Patients with COVID-19 show manifestations of respiratory tract infection, such as fever, cough, pneumonia, and in severe cases, death. Without a decisive mortality rate yet (currently estimated from 1–6%), COVID-19 is believed to be less deadly than SARS (~10%) or MERS (~40%), while its reproductive number has been estimated to be 2.0–6.57, which is higher than SARS and MERS (Petrosillo et al., 2020). As of November 23, 2020, COVID -19 has spread to other continents with multiple epicenters and caused 1,393,305 deaths out of 58,900,547 confirmed cases (World Health Organization (WHO), Coronavirus disease 2019 (COVID-19) Situation). Although certain physical treatment has been shown to assist patients to fight COVID-19 with their own immune systems, no proven remedies exist so far, inducing high mortality rates especially in senior groups (Zhou et al., 2020a). The development of effective prevention and treatment is an urgent need, particularly for life-threatening severe cases and to prevent or curb for future possible outbreaks. CAP disinfection of viruses such as SARS-CoV-2 (the virus responsible for COVID-19) is caused by plasma-generated reactive species inducing virus leakage and functionality loss. We recently demonstrated CAP inactivation of SARS-CoV-2 for a wide range of common surfaces (Chen et al., 2020f). Similar to other efforts toward CAP inactivation of viruses, our study found that the levels of these reactive species can be adjusted by plasma source design (e.g., electrode orientation and gas feed mechanism), feeding gas types, and operating conditions (e.g., electrode voltages and gas flow rates), and will vary additionally due to the nature of the product/substrate and the micro-organism. The results of this study also show the promise for SARS-CoV-2 inactivation in aerosols.

On the other hand, the importance of primary infection control measures has been highlighted for COVID-19. Clean hands frequently with alcohol-based hand rub or soap and water has been cited by the WHO as being "one of the most important hygiene measures in preventing the spread of infection." Even though the primary means of COVID-19 transmission is via aerosols, hands are one of the most frequent transmission routes for many infections because they come in direct contact with known portals of entry for pathogens, such as mouth, nose, and eyes. For such topical uses, CAP-treated liquid has increasingly been shown as an important approach in the field of disinfection. Free radicals and reactive species generated in the atmosphere and at the plasma-liquid interface can penetrate and enter the so-called "bulk medium" to react or recombine, giving rise to the generation of more stable ROS and RNS (such as H_2O_2 and NO_2-) (Bruggeman et al., 2016). According to the Centers for Disease Control (CDC), H_2O_2 is a stable and effective disinfectant against viruses. Thus, CAP-treated liquid can be used as a hand sanitizer to prevent the infection transmission Virus inactivation by CAP-treated liquid results from the combined action of a low pH and a high positive oxidation-reduction potential. CAP-treated liquid can also kill active virus on medical PPE, surfaces, instruments, etc. (Chen and Wirz, 2020). Such a CAP-based liquid can be of extreme value, extending the functions of sterilizing/disinfection for a many uses, including at home, in the office, public spaces, at entrances of buildings, in hospitals, and in laboratories. This technology can avoid the need for sanitizers that require consumables such as alcohol that are susceptible to supply chain disruption.

4.2   WOUND HEALING

Millions of people in the western world are affected by defective wound healing, which has a major burden on patients and healthcare systems (Schmidt et al., 2017a; Sen et al., 2009). Chronic wounds include, but are not limited to, pressure ulcers, diabetic foot ulcers, and venous leg ulcers (Frykberg and Banks, 2015). Cleaning chronic wounds to eliminate the excessive bacterial burden and necrotic tissue is the initial step and plays an important role in the treatment. Antimicrobial strategies are the second step, with the goal of removing or killing bacteria together with stimulating the wound's physical environment or the patient's general health (Daeschlein, 2013; Haertel et al., 2014). Although there are many wound therapies available, challenges remain in the treatment of more severe or complicated wounds (van Koppen and Hartmann, 2015). CAP can interact with cells and tissue leading to fast, painless, and efficient disinfection with great potential for wound healing (Arndt et al., 2013; Bernhardt et al., 2019). Today, there are several CAP-based medical devices (such as PlasmaDerm, kINPen MED®, and Adtec SteriPlas) on the market that are registered for the treatment of wounds and pathogen-associated skin diseases (Arndt et al., 2018; Brehmer et al., 2015; Weltmann et al., 2009). The first clinical trials were conducted by Isbary et al. using MicroPlaSter α® using Ar with a treatment time of 5 min and showed the efficacy and tolerability of CAP for infected chronic wounds (Isbary et al., 2010). Their follow-up study clearly elucidated that a 2-min treatment time of plasma was enough to achieve a strongly significant reduction of the bacterial load in chronic wounds in patients (Isbary et al., 2012). There are many clinical studies on wound healing having demonstrated the positive effects of plasma (Assadian et al., 2019; Bernhardt et al., 2019).

Figure 4.2 shows the CAP treatment of skin damage and wound healing. When the skin is injured, it displays an extraordinary ability for self-healing and to regenerate itself via a highly regulated biochemical cascade (Borena et al., 2015). Wound healing generally is categorized into four distinct phases: the hemostatic phase (clotting and vascular response), inflammatory phase, proliferative (repair) phase (scar formation, epithelial healing, and contraction), and the maturation phase (remodeling of tissues and epithelialization), as shown in Figure 4.2a. The inflammatory response can be regarded as the key process, as it is critical for cell proliferation, migration, and defense against bacterial contaminants (Lin et al., 2003). Stasis at this point could induce the destruction of the extracellular matrix, development of necrotic tissue, and persistent impairment of healing (Guo and DiPietro, 2010). Similarly, prolonging the inflammatory cycle prevents progression into the proliferative phase and results in the development of a self-healing or delayed state. The fibroblasts within self-healing wounds become unresponsive to cytokines and growth factors and take on a senescent form (Church et al., 2006; Lloyd et al., 2010). As a remedy, CAP has been shown to increase the proliferation of fibroblasts and other cells through plasma-generated RNS, to reduce the prolongation of the inflammatory phase of wound healing, and to reduce wound odors

(Stoffels et al., 2006). CAP also kills microorganisms, destroys bacterial spores, and controls biofilm generation without necessarily affecting mammalian cells. Thus, CAP is suitable for applications for a wide range of acute and chronic cutaneous wounds, especially with damage to the outer layers of the skin leading to infection and poor healing.

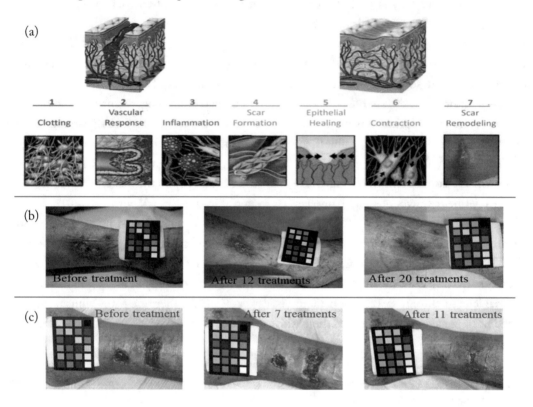

Figure 4.2: CAP for the repair of skin damage and wound healing. (a) The sequence of molecular and cellular events in skin wound healing. Reproduced from (Lloyd et al., 2010). (b) Inflamed ulcer of the right lower leg before treatment, after 12 treatments, and after 20 treatments (5×/week) with cold atmospheric Ar plasma generated by MicroPlaSter. Reproduced from (Heinlin et al., 2011). (c) Venous ulcers before plasma treatment, after 7 treatments, and after 11 treatments (2 min per day) by MicroPlaSter. Reproduced from (Heinlin et al., 2010). Reduction in the bacterial count and improved immunomodulatory effects of Ar plasma therapy may result in faster closure of chronic ulcers.

Figure 4.2b and c show clinical application of cold plasma on an inflamed ulcer and a venous ulcer (Heinlin et al., 2011, 2010). One clinical trial found that plasma treatment effected highly significant (34%, $P < 10^{-6}$) reduction in bacterial load in wounds compared with untreated wounds, a finding that is mirrored for all kinds of bacteria (Isbary et al., 2010). For diabetic foot disease, plasma-treated wounds healed to complete wound healing in half the time of untreated wounds in

48 patients while reducing pain (Heinlin et al., 2011; Isbary et al., 2014). These advances in tissue regeneration depend on RNS generated by cold plasma, which also plays an important role in wound healing in addition to stimulating regenerative processes. Some of the possible mechanisms by which CAP-generated RNS can stimulate wound healing are normalized microcirculation and vasodilatation, bactericidal effects, stimulating bacterial phagocytosis, inhibiting free oxygen radicals, clearing of necrotic detritus by neutrophils and macrophages, regulating immune-deficiencies, improving nerve conductance, boosting fibroblast and vascular growth by secretion of cytokines, affecting fibroblast proliferation, accelerating collagen synthesis, and increasing keratinocyte proliferation (Shekhter et al., 1998, 2005). Separately, ROS concentrations have been observed to be modulated during different healing phases, suggesting a physiological role of oxidants and antioxidants in wound healing and diabetes (Bekeschus et al., 2016; Fang, 2004; Houstis et al., 2006). This creates exciting possibilities for the therapeutic use of plasma-generated ROS in wound healing.

In experimental animals, plasma-treated wounds closed significantly faster when compared with the control group, and bacterial load was only observed in the control group (Shekhter et al., 1998). Erythrocyte slugging, micro thrombosis, and endothelial destruction were less expressed in treated animals, and wounds intentionally contaminated with *Staphylococcus aureus* had a shorter healing time by 31.6%. Kim et al. (2015c) characterized effects of Ar/Air plasma on wound healing, and found that plasma increased thickness of the epidermis, epidermal proliferation, and dermal extracellular matrix production, as shown in Figure 4.3a. Spinous and basal layers were especially thicker than in the control treatment group. Plasma did not induce structural deformation of the dermis or any physical damage to the stratum corneum layer. Similar results have been found in many other studies (García-Alcantara et al., 2013; Ngo Thi et al., 2014; Xu et al., 2015). CAP treatment also increases NO accumulation in the living tissue and stimulates NO synthesis through inducible nitric oxide synthase (Heuer et al., 2015; Kubinova et al., 2017; Shao et al., 2016). Inducible nitric oxide synthase, a highly diffusible intercellular signaling molecule, is an enzyme related to endogenous NO release and regulates many processes in skin physiology (Bruch-Gerharz et al., 1996; Xie et al., 1992). In studies, low NO bioavailability impaired wound repair, while high levels of inducible nitric oxide synthase accelerated wound repair in chronic leg ulcers (Luk et al., 2005; Pérez-Gómez et al., 2014). NO dosage in tissue is likely an important molecular mechanism responsible for plasma-induced acceleration of the healing process and wound closure. The efficacy of NO is assumed to be based on its functional influences on matrix deposition, cell proliferation, inflammation, angiogenesis, and remodeling (Luo and Chen, 2005). Moreover, Kubinova et al. indicated that macrophage infiltration into the wound area was not affected by the plasma treatment, indicating that the inflammatory phase was not disrupted during the healing process (Kubinova et al., 2017). No significant changes between plasma treatment group and control were found regarding the expression of genes associated with wound remodelings, such as angiogenesis (*Vefga*), collagen (*Col1a2*), matrix metalloproteinase (*Mmp2* and *Mmp14*), macrophage polarization (*Cd86*, *Cd163*), growth factors (*Egf, Fgf2, Pdgfa*), and cell proliferation (*Mki67*).

In addition, the potent cytoprotective effect of keratinocyte growth factor (*KGF*) on different types of epithelial cells has been demonstrated by a number of studies. In injured or inflamed tissues, KGF expression is strongly increased, which is important for the healing process (Frank et al., 1995). Nuclear E2-related factor (*Nrf2*), identified by Braun et al. as a novel target of KGF action, is reported as being regulated by growth factors on the transcriptional level (Braun et al., 2002). As one of the "cap-n-collar" transcription factors that control antioxidant proteins, transporters, and the encoding of phase II enzymes *Nrf2* participates in clearing toxic compounds and defense against antioxidants (Muzumdar et al., 2019). Figure 4.3b shows nuclear translocation of *Nrf2 in vitro* elucidated in dermal fibroblasts and epidermal keratinocytes, exemplifying plasma activation of the *Nrf2* pathway. Plasma treatment enhanced *Nrf2* expression and downstream targets, such as glutathione reductase, glutathione peroxidase 2, thioredoxin, and catalase: cytoplasmic superoxide dismutase was significantly decreased. Plasma treatment results into the upregulation of growth factor and angiogenic transcripts (e.g., *KGF, FGF, VEGFA, HBEGF, CSF2, bFGF, CD31, Akt/p-Ak*t) in human keratinocytes (Barton et al., 2013). Furthermore, *Nrf2* is a key regulator in innate immune cells (macrophages) (Parameswaran and Patial, 2010; Wakabayashi et al., 2010). Macrophages are secreted and respond to a repertoire of cytokines in a paracrine or autocrine manner, thus supporting the inflammatory response. Plasma treatment can increase neutrophil influx and macrophage presence, which upregulates pro-inflammatory markers *TNFα* or *IL-6* at early but not later stages of wound healing. Plasma treatment accelerated granulation and re-epithelialization in human and animal tests, supporting the importance of the role reactive species-mediated signaling plays in programmed death and cell proliferation (Fathollah et al., 2016; Gristina et al., 2009; Ulrich et al., 2015). Thus, plasma-derived RNS/ROS promoted wound closure, which was associated with increased proliferation and decreased apoptosis, an upregulated transcription of both anti-inflammatory and pro-inflammatory mediators, a spurred antioxidative defense response driven by *Nrf2*, an enhanced infiltration of macrophages and neutrophils, and enhanced angiogenesis (Schmidt et al., 2019). Moreover, CAP treatment is effective in killing bacteria without harmful effects on wound healing scenarios or promoting excessive scar formation. Treatment of wounded pig skin, which closely resembles human skin, by CAP did not induce any toxic effects on the skin, and effective/fast blood coagulation was observed (Dobrynin et al., 2011). Although plasma influenced human skin physiology parameters, no damage to the skin itself or skin functions was found (Boehm and Bourke, 2018; Fluhr et al., 2012; von Woedtke et al., 2019). It can thus be concluded that plasma treatment is safe for living intact and wounded skin at plasma doses several times higher than those required for the inactivation of bacteria.

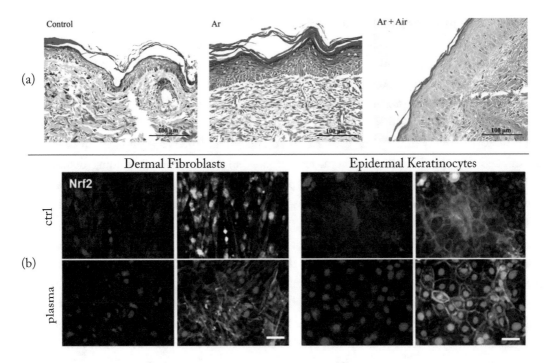

Figure 4.3: Acceleration of wound healing via plasma treatment. (a) Mice tissues treated with plasma and examined with Masson's Trichrome staining in case of control, Ar plasma, and Ar + Air plasma. Reproduced from (Kim et al., 2015c). (b) Nuclear E2-related factor (Nrf2) translocation from the cytoplasm (upper) to the nucleus (lower) after plasma treatment (60 s) as shown in dermal fibroblasts (left) and in epidermal keratinocytes (right). Reproduced from (Schmidt et al., 2019).

4.3   CANCER THERAPY

Cancer and cancer-related diseases remain among the leading causes of death and disability, accounting for approximately 13% of human death cases, and more than 7 million fatalities each year (Chen et al., 2017c). By 2030, annual cancer morbidity is projected to increase to 11 million deaths, even in light of the wide range of methods that are routinely being developed and employed for cancer treatment including traditional drugs, surgical techniques, radiation- and chemical-based therapies, immune therapies such as CAR-T cells, and many others. The wide range of cancer types and related complications demand that more progress and treatment methods are needed to cope with this persistent problem and reduce both the mortality rate and the long-term effects of treatment and survivorship. CAP has recently been shown to affect cells/tumors not only via direct treatment but also by indirect treatment (Dai et al., 2018; Dubuc et al., 2018; Keidar, 2018). Due to limitations in some situations, biomedical devices (such as needles, tubes, and microneedle patches) can assist CAP in treating tumors at restrictive positions (Chen et al., 2020a, 2017b; Mirpour et al., 2016). Plasma can also combine with other drugs/materials to be used as oral medication or injected into the blood for cancer therapy (Chen et al., 2016e; Li et al., 2019b). In addition, the effectiveness of plasma-treated media/liquids increases the potential for clinical applications of cancer therapy because it can be prepared in advance and stored until use (Harley et al., 2020; Zhou et al., 2020c). Clinical use of CAP is a relatively new field that grew from research in cancer therapies. CAP shows selectivity in targeting cancer cells and potentially offers a minimally invasive surgery option, allowing specific cells to be removed without influencing the surrounding tissue (Keidar, 2015; Schlegel et al., 2013). Various aspects of plasma treatment of cancer have been studied worldwide both *in vitro* and *in vivo*. We will discuss plasma for cancer therapy from *in vitro* to *in vivo* in this section.

Figure 4.4 shows the selectivity effect of plasma treatment. In one study, 60–70% of SW900 cancer cells were detached from the plate in the zone treated by plasma, whereas no detachment was observed in the treated zone for the Normal Human Bronchial Epithelial cells (NHBE) cells under the same treatment conditions (Figure 4.4a) (Keidar et al., 2011). CAP treatment resulted in a significant reduction in SW900 cell count, while NHBE cell count remained largely unchanged. In a separate study by Iseki et al. (2012), ovarian cancer cells (SKOV3) and healthy cells (WI-38) were treated with either Ar plasma or with Ar gas as a control (as shown in Figure 4.4b) . After 72-h plasma treatment, the proliferation rate of plasma-treated SKOV3 cells reduced to about 60% when the exposure time was 200 s/dish and less than 40% when the exposure time was 600 s/dish. More than 80% of WI-38 cells remained viable at the same plasma exposure of 200 s/dish, and WI-38 viability remained near 80% at 600 s/dish plasma exposure. Ar gas did not affect proliferation rates at any exposure. Figure 4.4c shows plasma-activated saline solutions have a stronger effect on cancer cells than that on healthy cells (Chen et al., 2017a).

One potential mechanism of plasma's selective targeting of cancer cells relates to the genetics of cancerous tissue. For example, a review by Luo et al. reported that the transformation of cells to a malignant state results from mutations that increase functionality in oncogenes and decrease the functionality of tumor suppressor genes (Raj et al., 2011). This in turn may induce the types of cell deregulation often observed as a result of enhanced cellular replicative, oxidative, proteotoxic, and metabolic stress, as well as damage to DNA. These functional changes may force cancer cells to reply on functionality very different from that of non-cancerous cells in order to maintain vitality, and targeting these genes can lead to selective death of cancer cells as demonstrated by Yap et al. (2011). CAP may be one method to target these transformed non-oncogenes and, by disrupting their function, may induce selective death; such induced death may be either necrotic, in which the change itself is lethal to the cell, or through the initiation of cellular self-destruction via apoptosis. It affects these non-oncogene dependencies in the context of a transformed genotype and might induce a synthetic lethal interaction and the selective death of BxPC-3 cancer cells compared to H6c7 normal cells (as shown in Figure 4.4c).

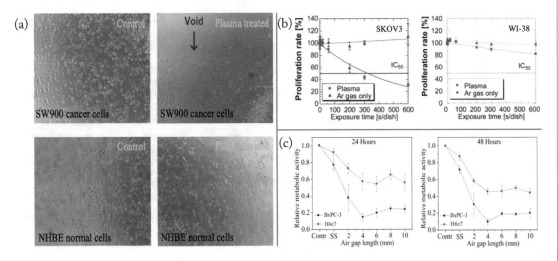

Figure 4.4: The selectivity of plasma treatment. (a) SW900 cancer cells were detached from the plate in the zone treated with plasma, whereas no detachment was observed in the untreated zone for the normal NHBE cells. Reproduced from (Keidar et al., 2011). (b) Ovarian cancer (SKOV3) cells and fibroblast (WI-38) cells were treated with plasma for 2, 6, 20, 60, 200, 300, and 600 s/dish. 72 h after plasma treatment, cell proliferation was evaluated by MTS assays. Proliferation rates were calculated by normalizing to non-treated cells. As a control, Ar gas was applied at the same flow rate (2 SLM (standard liter/min)) as the plasma treatments. Reproduced from (Iseki et al., 2012). (c) Effect of 7 media: RPMI/KSFM (cultured media), saline solution (SS), and 5 plasma-activated media (SS activated by plasma for 40 s at 2, 4, 6, 8, and 10 mm distance) on the viability of BxPC-3 human pancreas cancer cells and H6c7 human pancreas normal cells after 24 h and 48 h incubation, respectively. The percentages of surviving cells for each cell line were calculated relative to controls (RPMI/KSFM). Reproduced from (Chen et al., 2017a).

Other mechanisms involving reactive species also exist. H_2O_2 reacts with NO_{2-} to form peroxynitrite $OONO^-$ and H_2O (Tian and Kushner, 2014). $OONO^-$ is a powerful oxidant and nitrating agent that is known to be more damaging to cancer cells than non-cancerous cells (Beckman and Koppenol, 1996). Separately, intracellular ROS-mediated up-regulation of gene Death Receptor 5 (*DR5*) has been shown to result in apoptosis, and intracellular generation of ROS induces increased expression of proapoptotic protein *Bax*, disruption of the mitochondrial membrane potential and release of cytochrome c and AIF into the cytosol resulting in apoptosis via the activation of caspase-3 cascade (Kong et al., 2012; Zhang et al., 2008). Therefore, plasma-induced apoptosis in cancer cells might be orchestrated by the synergistic effects of ROS on both extrinsic and intrinsic pathways.

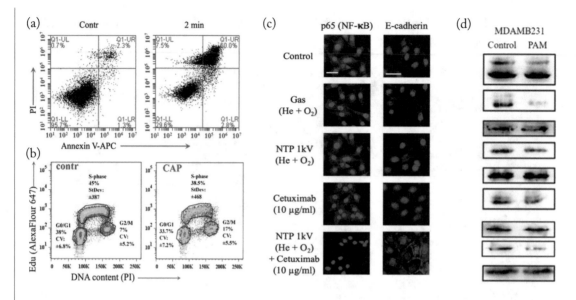

Figure 4.5: Plasma treatment *in vitro*. (a) Cell apoptosis analyzed by flow cytometry after He + 0.5% O_2 plasma treatment for 2 min. Reproduced from (Xu et al., 2018a). (b) Detailed visual of cell cycles for control and 60-s CAP treatment, with the correlation between DNA content and DNA-replicating cells shown for carcinoma cells (PAM212) after 24 h. Reproduced from (Volotskova et al., 2012). (c) Immunocytochemical assay for p65 (NF-κ B) and E-cadherin for various treatment combinations of gas, cold plasma, and cetuximab. Scale bar = 50 μm. Reproduced from (Chang et al., 2015). (d) Total and phosphorylation levels of JNK, p38, ERK, p65 in MDA-MB-231 before and after PAM treatment at 8 h. Reproduced from (Xiang et al., 2018).

A large number of studies have demonstrated CAP initiates cytotoxic activities in various cancer cell lines, by either direct exposure to plasma or indirect exposure via plasma-activated medium (PAM) prepared prior to cell treatment. PAM offers the advantage of being produced beforehand as a "drug," making it more convenient to target cancer cells or tumors (Chauvin et al., 2019; Chen et al., 2016e; Freund et al., 2019; Liedtke et al., 2017; Utsumi et al., 2013). As a complement to direct plasma, PAM can be stably stocked for several days and target areas for which direct exposure is unavailable. CAP and PAM can effectively kill tumor cells or tissues via reactive species generated by the plasma itself or through plasma/medium interactions. Figure 4.5a indicates apoptosis detected by flow cytometry 24-hour after He + 0.5% O_2 plasma treatment (Xu et al., 2018a). The authors indicated that ROS generated by plasma could lead to cell apoptosis through activation of protein CD95 and the subsequent downstream caspase cascade. Figure 4.5b shows the assessment of cell counts in the three distinct phases of the cell cycle (G0/G1, S, and G2/M) (Volotskova et al., 2012). Compared to the control, plasma-treated cells showed decreases in the percentage of cells in G0/G1 and S phase, with a corresponding increase visible for the G2/M phase.

This implies that more cells are in a paused state in G2/M rather than dividing and reproducing, and thus that plasma treatment effectively reduced the growth rate of the cells in a phase-specific way. Figure 4.5c presents the combination of plasma treatment with growth-receptor-inhibitor cetuximab to be effective in regulating the protein expression of *NF-κB* and cell adhesion molecule E-cadherin markers in human oral cancer (SCCQLL1) cells (Chang et al., 2015). The transcription factor *NF-κB* is an important protein, whose dysregulation is implicated in various cancers, with disruption of *NF-κB* pathways as acritical characteristics of high-risk head and neck cancer in particular (Chung et al., 2006). Although gas-, plasma-, or cetuximab-only treatment did not significantly affect *NF-κB* expression, little *NF-κB* was detected in SCCQLL1 cells in the plasma/cetuximab combination group. In most cancers, *NF-κB* is involved in multiple stages of tumor growth, including tumorigenesis, progression, and maintenance, in addition to resistance to cytotoxic chemotherapy implicating the improper expression of *NF-κB* in malignancy and resistance to therapy (Tanaka et al., 2011). As a molecule associated with cell adhesion, E-cadherin (normally observed at the cell membrane) is expressed in lower quantities for cancer cells, while its expression was increased after plasma/cetuximab combination therapy in SCCQLL1 cells. In another study, Xiang et al. employed western blot to investigate the total level and phosphorylation status of four mitogen-activated protein kinases (*JNK, ERK, p38,* and *p65*), which are enzymes involved in cellular reproduction and stress responses, in triple negative breast cancer cells (MDA-MB-231) after PAM treatment for 8 h (Xiang et al., 2018). In general, boosted levels of these enzymes are found in breast cancers. PAM was shown to significantly inhibit the phosphorylation levels of JNK and *p65* (*NF-κB*) but has little effect on *p38* and *ERK* (as shown in Figure 4.5d), inhibiting the growth and migration of MDA-MB-231cells. In addition, oxidative stress elicits *JNK* phosphorylation in general and inhibits *JNK* phosphorylation in cancer cells (Peng et al., 2018c). As PAM increased oxidative stress and suppressed cancer cell proliferation, cells with higher proliferation ability are likely to be more sensitive to plasma treatment (Naciri et al., 2014).

CAP induces a significant reduction of tumor size and a corresponding lengthening of survival time. Figure 4.6 shows several aspects of plasma treatment *in vivo*: plasma delivery penetrating the skin to reach the tumor (Figure 4.6a) and plasma delivery direct to the tumor (Figure 4.6b). Keidar et al. (2011) applied a cold plasma jet to nude mice bearing subcutaneous bladder cancer tumors (Figure 4.6a). A single transcutaneous plasma treatment resulted in ablation of the tumor through the overlying skin. Tumors of ~5 mm in diameter were ablated following single 2-min plasma treatment, with no damage to the skin observed from treatment times of 2–5 min. Tumor growth rates were markedly decreased after plasma treatment, although the tumors eventually recurred. Notably, plasma treatment induced a markedly improved survival in the treatment group, with a median survival duration of 33.5 vs. 24.5 days for the control group. For *in vivo* treatment, Chen et al. (2017b) developed a novel micro-sized CAP device (*μCAP*) and employed it to target glioblastoma in the murine brain. He *μμCAP* was directly applied for 30 s to tumors in the brain of living mice via an implanted endoscopic tube as shown in Figure 4.6b. In the control animals, there was a clear increase in tumor volume over the 2-day period. In contrast, He μCAP-treated animals fell below baseline values. Employing *in vivo* bioluminescence imaging, the tumor volume in a control animal (helium only) increased by approximately 600%, whereas He μCAP-treated tumor volume decreased by approximately 50% compared with baseline levels (Chen et al., 2018c). These striking findings indicate the potential of μCAP to inhibit glioblastoma growth *in vivo*.

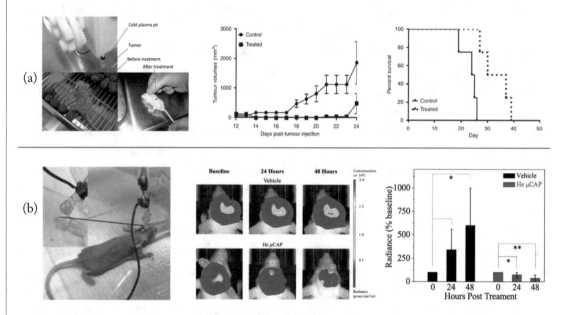

Figure 4.6: CAP treatment *in vivo*. (a) Transepidermal plasma treatment of tumors: the typical image of mice with a single tumor before and approximately 1 week after treatment; plasma treatment effect on the growth of established tumors in a murine melanoma model; and plasma treatment effect on the mice survival in a murine melanoma model. Reproduced from (Keidar et al., 2011). (b) Targeting of glioblastoma tumors with direct application He μCAP: photograph of a new μCAP device for plasma delivery through an intracranial endoscopic tube to target glioblastoma tumors in the mouse brain; representative *in vivo* bioluminescence images illustrating glioblastoma tumor volume (i.e., light emission or radiance) at baseline and 2 days following He μCAP or vehicle (helium) treatment. Areas of high photon emission are shown in red and low are shown in blue; group summary data (n = 3/group) indicates that the tumor volume in control aggressively increased following treatment, whereas He μCAP delivery-maintained tumor volume below basal levels. Reproduced from (Chen et al., 2017b).

The efficacy of plasma in targeting tumors is primarily due to plasma containing energetic ions, free radicals, reactive species, UV radiation, and the transient electric fields inherent to plasma delivery, all of which interact with cells and other living organisms (Semmler et al., 2020). Radicals and electrons generated during plasma formation can be either short lived or long lived. If these radicals or electrons reach a solution, they can be involved in many complex reactions that result in the formation of other short- and long-lived radicals or species. Short-lived radicals or species include superoxide (O_{2-}), nitrite (NO), atomic oxygen (O), ozone (O_3), hydroxyl radical (•OH), singlet delta oxygen (SOD, $O_2(^1\Delta g)$), peroxynitrite ($ONOO^-$), and many more (Graves, 2012). Long-lived species include hydrogen peroxide (H_2O_2) and nitrite (NO_{2-}) (Tian and Kushner, 2014). Short- and long-lived species or radicals in mice brains including •OH, NO, O_{2-}, O, O_3, $O_2(^1\Delta g)$, $ONOO^-$, H_2O_2, and NO_{2-} are prominent components of antitumor plasma action (Dobrynin et al., 2009; Xia et al., 2019). O (including the ground state and all the excited states) is known to play multiple important roles in cells (Walsh and Kong, 2008). Plasma also generates a significant amount of O_3, which is believed to have a strongly aggressive effect on cells (Fridman et al., 2005). $O_2(^1\Delta g)$ is another important ROS with an excitation energy of 0.98 eV. Highly reactive molecule $O_2(^1\Delta g)$ not only produces oxidative damage in many biological targets but is also a primary active species in the selective killing of tumor cells in emerging cancer therapy (Dougherty et al., 1998; Schweitzer and Schmidt, 2003). O_{2-} generated by CAP can activate mitochondrial-mediated apoptosis by changing the mitochondrial membrane potential and simultaneously up-regulates pro-apoptotic genes and down-regulates anti-apoptotic genes for activation of caspases-driven cell death (Riedl and Shi, 2004). •OH_- derived amino acid peroxides lead to cell injury because •OH itself and protein (amino acid) peroxides are able to react with DNA, thereby resulting into various forms of genetic damage (Adachi et al., 2015; Gebicki and Gebicki, 1999). In addition, NO_{2-} and O_{2-} can easily react to form $ONOO^-$, which is a powerful oxidant and nitrating agent known to be highly damaging to tumor cells (Pacher et al., 2007).

The specifics of plasma interaction with cells and tumors are still somewhat elusive. The current understanding of the primary mechanism of plasma anti-cancer treatment is centered on ROS and RNS generation. In general, computational models (Figure 4.7a) and experiments (Figure 4.7b and c) are employed to help understand plasma's anti-cancer mechanisms. Xu et al. (2018b) employed reactive molecular dynamics simulations to study the interaction mechanisms between plasma-generated species and various tissues, and proposed a potential mechanism of plasma anti-tumor action as shown in Figure 4.7a. Plasma-generated ROS triggered a "bystander effect" based on gap junction intercellular communication, which can trigger apoptosis in cells not directly treated and has been shown to target tumor cells more easily. ROS and RNS move into the bystander cell to induce cell death via the bystander effect, and then move in the next cell, continuing the process.

This important process is indicative of the many ways that CAP can facilitate advantageous behaviors in complex systems. The current understanding is that •OH and $HO_2^•$ (perhydroxyl radical) radicals generated by plasma can chemically react with the N-terminal of gap junction resulting in structural damage. There are two identified breaking mechanisms: C-N peptide bonds and C-C bonds can be damaged by •OH and $HO_2^•$ radicals, respectively. The interactions between proton leak (a branch pathway in the utilization of proton-motive force to produce heat) and electron leak (a branch of electron transfer directly from the respiratory chain to form $O^{\bullet}_{2-}$) happen in mitochondria via consumption of H^+ by $O^{\bullet}_{2-}$ anions to form a protonated $HO^{\bullet}_2$ (Liu, 1997). $HO^{\bullet}_2$ is permeable across the mitochondria membrane and results into the leakage of high-energy protons of proton-motive force. In addition, the ΔpH component of proton-motive force may provide a suitable chemical environment, with the H^+ from the redox proton pumps could be mainly localized on the inner surface of the mitochondria membrane. There, $O^{\bullet}_{2-}$ may consume the protons and conduct them across the membrane to the matrix side via a pK-dependent formation of the protonated species of $HO^{\bullet}_2$ (Bielski et al., 1985). The endoplasmic reticulum (ER) redox environment dictates the fate of entering proteins, and the presence of redox signaling mediators regulates the abundance of ROS (Zeeshan et al., 2016). In the pathway of oxidative proteins, electrons transfer from protein disulfide isomerase (PDI) to endoplasmic reticulum oxidoreductin-1 (ERO-1) and molecular oxygen via a Flavin adenine dinucleotide (FAD)-dependent reaction. It suggests that ERO1 has a strong association with protein load in the ER and can trigger ROS generation. In the presence of transition metals (such as Fe^{2+} or Cu^{2+}), H_2O_2 can be converted to $\bullet OH$, which is highly reactive and causes damage to lipids, proteins, DNA, and gap junctions (GJ). $\bullet NO$ is a reactive radical produced from arginine by nitric oxide synthase (NOS) (Förstermann and Sessa, 2012). $\bullet NO$ has a very short half-life and can react with superoxide to form peroxynitrite ($ONOO^-$), which can modify the structure and function of proteins and, as stated previously, is highly damaging to tumor cells. Figure 4.7b shows the signaling mechanism for PAM-induced A549 cell injury via a spiral apoptotic cascade involving the mitochondrial-nuclear network (Adachi et al., 2015). The diffusion of H_2O_2 from the extracellular space as well as its derived ROS could trigger mitochondrial dysfunction as the primary event in the signaling pathway. The *Bcl2* family is a key regulator of the mitochondrial pathway, which is one of the most important apoptosis signaling pathways in cell (Lindsay et al., 2011). *Bcl2* prevents apoptosis by reducing Ca^{2+} levels in damaged ER (Wegierski et al., 2009). The activation of TRPM2 is believed to induce an elevation in $[Ca^{2+}]_i$ through the influx of extracellular Ca^{2+} and release of lysosomal Ca^{2+} (Ru and Yao, 2014). In addition, reactions with DNA induced by ROS activate *PARP-1*, which catalyzes the cleavage of NAD^+ into nicotinamide and ADP-ribose to form large amounts of polyADPR. This reaction leads to the consumption of NAD^+ and depletion of ATP. In turn, reductions in $\Delta\psi m$ increase the permeabilization of the mitochondrial membrane and allow the efflux of AIF. In Figure 4.7b, H_2O_2 and/ or its derived ROS may disturb the mitochondrial-nuclear network by reducing $\Delta\psi m$, decreasing the expression ratio of *Bcl2/Bax*, activating *PARP-1*, releasing AIF, and depleting NAD^+, with a resulting increase in ER stress. The accumulation of ADPR as a product of the above reaction as well as extracellular/intracellular H_2O_2 activates TRPM2, which causes the extracellular influx of Ca^{2+} as well as its release from intracellular stores.

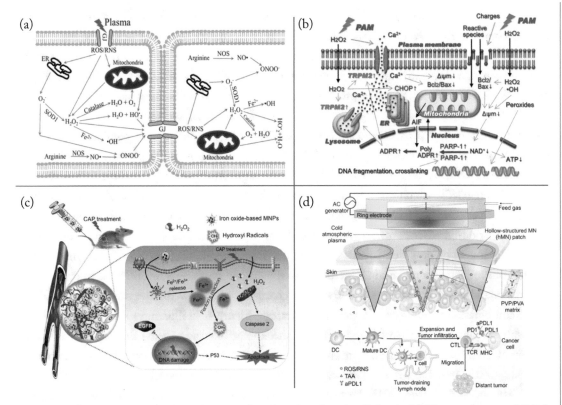

Figure 4.7: Potential signaling mechanisms for plasma-induced cell death. (a) Plasma-generated ROS induced bystander effect via gap junction (GJ): cell death induced by plasma-oxidative stress in neighboring cells is mediated by a bystander effect. In the left cell, stimulation of the mitochondria and endoplasmic reticulum (ER) via plasma-generated ROS leads to the generation of additional ROS. The transfer of ROS and RNS to neighboring cells via dysfunction GJ. Reproduced from (Xu et al., 2018b). (b) ROS in PAM diffuses into cells and disturbs the mitochondrial-nuclear network. Damage to the plasma membrane by ROS-cooperating charged species may not only induce apoptosis, but also increase permeability to other extracellular reactive species. Reproduced from (Adachi et al., 2015). (c) Magnetic nanoparticles enhancing tumor-selective killing effect of CAP. Plasma-generated reactive species will cause a noticeable rise of intracellular H_2O_2. Fe^{2+}/Fe^{3+} released from the lysosome containing magnetic nanoparticles could catalyzed H_2O_2 into •OH, which causes the damage to cancer cells, such as inducing mitochondria-mediated apoptosis and double-strand DNA breaks. Reproduced from (Li et al., 2019b). (d) Illustration of transdermal cold atmospheric plasma-mediated immune checkpoint blockade (ICB) therapy. Schematic of transdermal combinational CAP and ICB therapy assisted by the polymeric hollow-structured microneedle patch loaded with aPDL1. Nomenclature: TAA, tumor-associated antigen; CTL, cytotoxic T lymphocyte; TCR, T-cell receptor; MHC, major histocompatibility complex. Reproduced from (Chen et al., 2020a).

Figure 4.7c shows iron oxide-based magnetic nanoparticles (MNPs) enhancing the tumor-selective killing effect of CAP. CAP and iron oxide-based MNPs either individually or in combination significantly reduced mRNA and protein expression levels of *EGFR*. *EGFR* and its downstream signaling pathway are involved in the proliferation of cancer, as the activation of *AKT* and *ERK1/2* are also participate in proliferation and differentiation (Lau et al., 2017). Expression of *EGFR* in the cancer cell is regulated by *TGF-α*, E-cadherin, and some miRNAs. Signaling pathways of *EGFR-PI3K/AKT* and *EGFR-RAS/ERK* are important for the regulation of cellular proliferation and are implicated in tumor initiation and development (Balakrishnan et al., 2017). Co-treatment of CAP and iron oxide-based MNPs significantly decreased the expression levels of *pERK* and *pAKT*, which are important phosphokinases in tumorigenesis and tumor progression (Lau et al., 2017). H_2O_2 was generated in CAP-treated A549 cancer cells, while Fe^{2+}/Fe^{3+} from iron oxide-based MNPs can transform H_2O_2 into •OH through Fenton's reaction (Wang et al., 2011). Iron oxide-based MNPs can enhance the tumor-killing effect of CAP through this mechanism inducing mitochondria-mediated apoptosis and double-strand DNA breaking. A better understanding of the precise underlying mechanisms will allow for the determination of the "best" combination when used as a treatment strategy. Figure 4.7d illustrates the transdermal cold atmospheric plasma (CAP)-mediated immune checkpoint blockade (ICB) therapy for melanoma treatment. ICB increases antitumor immunity by inhibiting intrinsic down-regulators of immunity and has greatly transformed the landscape of human cancer therapeutics (Ribas and Wolchok, 2018; Sharma and Allison, 2015; Ye et al., 2018). Chen et al. (2020a) explored the potential of microneedle (MN)-array patch-based transdermal delivery that combined CAP and ICB (Figure 4.7d) . They leveraged the hollow-structured micro-needle patch as microchannels to allow CAP to be delivered through the skin into tumors. Cancer cell death induced by CAP released tumor-associated antigens and promoted dendritic cell (DC) maturation in the tumor-draining lymph node, where DCs could present the major histocompatibility complex- peptide to T cells. Subsequent T cell-mediated immune response is initiated and can be further augmented by a checkpoint inhibitor, such as anti-programmed death-ligand 1 antibody (aPDL1), included in the hollow-structured micro-needle patches. The synergism between CAP and ICB provided a broad platform to promote tumor elimination.

Thus, CAP is a great source of ROS and RNS and can modulate a number of intracellular/extracellular signaling pathways to induce cancer cells apoptosis, including the damage of cell membranes, nucleoids, proteins, enzymes, etc. (Bauer, 2016; Mitra et al., 2019). The ability of ROS/RNS to enter the cell and therefore interfere in these pathways is heavily influence by tumor-specific factors such as increased presence of cholesterol or aquaporins in the cell membrane as well as the cell's reduced protection against oxidative stress (Semmler et al., 2020). Although there are several known apoptosis pathways by which plasma kills tumor cells, more work still needs to be done to identify additional pathways and determine the pathways that are most attractive for cancer treatment. Plasma-induced biochemistry and natural killer cell anti-cancer mechanisms provide a chance to establish a rationale for plasma's selective killing of cancer cells in both *in vitro* and *in vivo* (Bauer and Graves, 2016; Gauthier et al., 2019; Graves, 2012). Plasma acts selectively against cancer cells and tumors at least partly due to its reactive species inducing self-perpetuating apoptotic signaling via cell-to-cell communication. The plasma-induced cell apoptosis response is communicated to bystander cells via intercellular communications (gap junction) and the release of soluble factors, including ROS/RNS, cytokines, or products of macromolecular species (Graves, 2014b; Xu et al., 2018b). In addition, plasma treatment induces the release of tumor-associated antigens, which promotes the maturation of DCs to initiate immune response (Chen et al., 2020a; Khalili et al., 2019; Lin et al., 2017, 2018a). Altered microenvironments are not just a protective element in tumors but in microbial biofilms as well, and CAP treatments serve as a potential method for helping to reverse many aspects of these altered microenvironments in both biofilms and tumors (Bekeschus et al., 2018; Gilmore et al., 2018; Privat-Maldonado et al., 2019a). The evidence to date suggests that CAP has a significant effect on several tumors in preclinical and clinical applications (Metelmann et al., 2015, 2018a; Schuster et al., 2016).

4.4   OTHER BIOMEDICAL APPLICATIONS

In addition to wound healing and cancer as discussed above, plasma works for dentistry, cosmetics, blood coagulation, dermatology, gynecology, tissue regeneration, and so on (Busco et al., 2019; Friedman et al., 2017; Kaemling et al., 2005; Liu et al., 2020b; Ly et al., 2018; Naujokat et al., 2019; Nejat et al., 2019; Weiss and Stope, 2018; Wirtz et al., 2018). Some plasma biomedical applications are shown in Figure 4.8. For example, a murine liver incision model was proposed to evaluate the efficacy of plasma-mediated hemostasis as shown in Figure 4.8a (Bekeschus et al., 2017). The cotton swab quantified blood loss by collecting excess blood. Plasma treatment appeared to induce a faster hemostatic response in animals treated with rivaroxaban or without anticoagulant treatment but not those treated with clopidogrel. The study authors thus demonstrated that CAP can induce a platelet-dependent pro-coagulative response *in vivo* during surgical procedures. Plasma-derived ROS/RNS oxidize platelets, thereby increasing their activation and mediating hemostasis. There may be three related factors: (1) blood with strong antioxidant capacity; (2) excessive stimulation driving platelets to undergo apoptosis; and (3) byproducts of aggregation of other cell types (Bocci et al., 1998; Kim et al., 2016b; Zharikov and Shiva, 2013). CAP is also as an innovative new medical technique for the treatment of Hailey-Hailey disease (Isbary et al., 2011). In this study, the addition of cold plasma therapy to the application of topical corticosteroids and local disinfectants induced a rapid improvement in the axilla, whereas the groin, treated only with topical corticosteroids and disinfectant, remained unchanged. The beneficial effect of cold plasma in Hailey-Hailey disease was likely partly based on plasma's bactericidal and fungicidal effects but also due to a positive effect of the oxidant-antioxidant system on skin lesions (Cialfi et al., 2010). ROS accumulation in keratino-cytes derived from cutaneous lesions implied that disturbances in the oxidant-antioxidant system might importantly affect the pathogenesis of this disease.

Figure 4.8b schematically illustrates the novel plasma jet application for peri-implantitis treatment. Two important basic conditions are required for patients with peri-implantitis: removing biofilms from implant surfaces and promoting osteoblast adhesion and spreading (Duske et al., 2015; Schwarz et al., 2018). Currently, there are several types of CAP being employed to remove oral biofilms suggesting plasma as a feasible option for the treatment of peri-implantitis (Shi et al., 2015; Ulu et al., 2018; Vogelsang et al., 2010). The plasma jet not only achieves much higher electron density but also produces greater quantities of ROS and RNS (Meemusaw and Magara-phan, 2013). Porphyromonas gingivalis was completely eliminated and different osteoblast cell lines were well attached to the modified titanium surface. Treatment via plasma jet improved the surface hydrophilicity and roughness of the implants, which is promising for the second condition of periimplantitis treatment.

Plasma application on human oral mucosa modulates tissue regeneration, as shown in Fig-ure 4.8c (Hasse et al., 2014). No thermal damage, the formation of blisters, or other skin damage could be detected after plasma treatment. Plasma treatment increased the percentage of (antibody) $\gamma H2A.X$-positive cells with no concomitant with the number of apoptotic cells. The concentration of repair proteins near $\gamma H2A.X$ foci was elevated (Kouzarides, 2007), and most TUNEL-positive (apoptotic) cells were detected after only a brief plasma treatment and did not increase substantially with longer treatment. Figure 4.8d shows expected improvements in periorbital wrinkles and laxity in cheeks, full face, jowls, and perioral regions after plasma treatment. Analysis of the degree of skin tightening indicated patients had a 70–80% reduction in skin lines at 3-month follow-up visits (Bogle et al., 2007). Histological studies performed on plasma-resurfacing patients have confirmed continued collagen production, reduction of elastosis, and progressive skin rejuvenation beyond one year after treatment. The U.S. Food and Drug Administration has granted 510 (k) clearance for plasma skin regeneration for treatment of rhytids of the body, actinic keratoses, seborrheic keratoses, superficial skin lesions, and viral papillomata. CAP offers a unique form of non-ablative resurfacing and is not chromophore dependent without inducing vaporization of the epidermis, leaving a layer of intact desiccated epidermis acting as a natural biologic dressing and promoting rapid healing (Foster et al., 2008). Beyond those already mentioned above, there are many more clinical applications of CAP that are in the testing or under development across a broad range of specializations (Chen et al., 2016f; Chesnokova et al., 2003; Fridman et al., 2008; Golubovskii et al., 2004; Seeger, 2005; Shekhter et al., 2005; Shulutko et al., 2004).

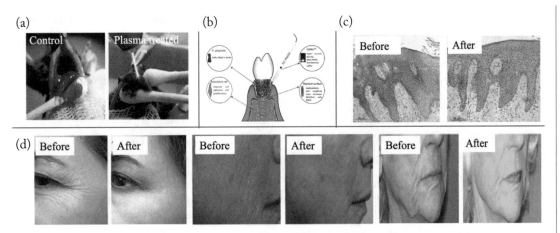

Figure 4.8: CAP for several biomedical applications. (a) A murine liver incision model: murine livers were incised resulting into blood loss quantified spectroscopically from blood taken up by cotton buds (control) and blood coagulation (plasma-treated). Reproduced from (Bekeschus et al., 2017). (b) Schematic illustration of the application of a novel plasma jet for peri-implantitis treatment. Reproduced from (Yang et al., 2018b). (c) Representative histological photographs of human oral mucosa without plasma treatment and after 5 min plasma treatment. No plasma-induced damage of epithelial layers and underlying lamina propria was detected. Magnification 100×. Scale bar represents 50 μm. Reproduced from (Hasse et al., 2014). (d) Plasma for skin regeneration and cosmetology. Before and after photos showing improvement on periorbital rhytids (1 year), cheek (30 days), and full-face with tightening of the jowls (30 days), in left, middle, right, respectively. Reproduced from (Foster et al., 2008).

One of the important species generated by CAP is nitric oxide (NO). NO is a gas that enables diverse biological activities and can interact with superoxide (O_2-), forming peroxynitrite (ONOO-) to mediate bactericidal or cytotoxic reactions in turn (Foresti et al., 1997). Also, NO is unstable and reacts with oxygen to form oxides of nitrogen (NO_x) (Chen et al., 2016d). Moreover, NO has played a major role in regulating airway function and in treating inflammatory airway diseases (Ricciardolo et al., 2004). The beneficial effects of NO inhalation are observed in most patients with severe ARDS (acute respiratory distress syndrome), since it impedes the synthesis of viral protein and RNA (Guo et al., 1995; Lei et al., 2020). In addition, organic NO donor, S-nitro-so-N-acetyl penicillamine, can remarkably obstruct the replication cycle of SARS-CoV (Åkerström et al., 2009; Keyaerts et al., 2004). NO inhalation has been shown as a potential treatment for lung cancer (Martel et al., 2020). It also has the potential to be a treatment of severe COVID-19. For example, on March 29, 2020, *Science News* reported a team led by Jalaledin Ghanavi had just completed a safety trial in 14 COVID-19 patients, half of whom received ionized helium with noticeable improvement. NO is most prominent for CAP devices that use nitrogen and oxygen as a mix feeding gas. Operators can adjust feeding gas mix ratio, operating conditions, and other parameters in a nitrogen atmospheric environment to reduce the presence of other oxides of nitrogen (such as

NO_2). Although patients with SARS inhale purified NO, NO reacts with O_2 to form NO_x in the trachea from mouth to lungs. However, it proves to remain effective, possibly because the toxicity of NO_x can be neglected compared with the seriousness of the COVID-19 disease. In all, CAP is a promising source for dose-controllable NO treatment of severe COVID-19 patients. In addition, SARS-CoV-2 employs multiple strategies to avoid immune responses to better survive in host cells. COVID-19-infected pneumonia patients typically have severe immune abnormalities and risk of cytokine release syndrome (CRS), which results into a decrease of T cells and Natural Killer (NK) cells and an increase of interleukin 6 (IL-6), CD4/CD8 ratio (CD: cluster of differentiation), fever, tissue/organ dysfunction, and an abnormal coagulation function (Ahmed et al., 2020; Wang et al., 2020a; Zheng et al., 2020). Currently, immunotherapy has been proved to an effective treatment to fight COVID-19 diseases. The potential mechanism is that inflammatory cells such as effector T cells and macrophages accumulate rapidly from peripheral blood by chemokines and release a large number of cytokines into the blood when they kill tumor cells, viruses, or bacteria (Sun et al., 2009). As another vector, stem cells are hardwired to express antiviral interferon-stimulated genes (Wu et al., 2018). In addition, β-glucan is capable of training both hematopoietic stem cells (HSCs) and myeloid progenitors (Maguire, 2020). These conditioned HSCs and myeloid progenitors can more efficiently ward off inflammatory challenges. With regards to these mechanisms and pathways, CAP has shown its great potential toward immunotherapy in biomedicine (Bekeschus et al., 2020a). Figure 4.7d shows that CAP treatment promotes dendritic cell (DC) maturation in the lymph node, where DCs can present the major histocompatibility complex-peptide to T cells. The subsequent T cell-mediated immune response can be augmented by the immune checkpoint inhibitors, resulting in enhanced local and systemic antiviral immunity (Chen et al., 2020a). Therefore, the adoptive therapy of He CAP should be an ideal choice to be employed for COVID-19 patients and similar infections.

4.5   SUMMARY AND FUTURE DIRECTIONS

In summary, CAP's ability to generate ROS and RNS makes it a prime source for future innovative cancer therapies through this role, but additional research into controlling and driving specific reactions is needed before sufficient understanding of the process exists for medical use. For example, the role of reactive oxygen species in causing a so-called "bystander effect" of cascading apoptosis in cancer tumors has been generally observed but little understood until recently (Xu et al., 2018b).

CAP shows potential as an intervention for a wide range of conditions, the complicated nature of its application will pose challenges in obtaining clearances from regulatory agencies. Even with these challenges, there are already several devices certified for clinical use and being sold to doctors, such as PlasmaDerm of Cinogy, kINPen® MMED of neoplas tool, or Plasma ONE of Plasma Medical systems (certified already for several years in Europe) (Arndt et al., 2018; Brehmer et al., 2015; Isbary et al., 2010; Weltmann et al., 2009). Tens of thousands of patients have already been treated with CAPs. Additionally, the Ar plasma coagulation has been used for almost three decades. CAP is, in effect, a "medical device" with pharmacological effects. An analogy might be something like an insulin pump, which has physical device components in addition to the medication administered; with CAP, the "device" component would be the mechanism by which plasma is generated, while the "medication" would be the plasma discharge and reactive species (Graves, 2012; Lu et al., 2019). The approval process will likely require multiple phases focusing separately on plasma and its constituent reactive species as delivered to the host and, independently, on the safe generation of plasma by the device. Further, the variation in results for different carrier gases, flow rates, frequencies, and generation methods will likely require either the creation of universal principles for plasma generation or separate approval for each implementation. As discussed at the end of Chapter 2, such commonality may generally help formalize research and development efforts in the overall CAP community.

The toxicity of cold plasma to healthy tissues as well as cancerous tissues needs to be more fully understood. While the chemotherapies and radiation treatments used currently against many cancers are themselves highly toxic, the specific nature of such toxicities are well known and constantly being updated (Green et al., 2001; Meirow et al., 2010). In contrast, the specific limits within which CAP can be toxic to cancers but nontoxic or minimally toxic to healthy tissues are still being researched. Other CAP-mediated cancer therapies may exist but will likely provide their own complexities; for example, targeted CAP-derived immunotherapies may help reduce toxicity but raise questions regarding the penetrative capability or deliverability of such treatments as well as the potential for unintended systemic responses.

CHAPTER 5

Plasma Environment

The natural environment encompasses the interaction of all living species, climate, and natural resources that affect, among other things, human survival, and economic activity (Johnson et al., 1997). Generally, the term natural environment evokes environmentalism (a broad political, social, and philosophical movement), which advocates various actions and policies in the interest of protecting what nature remains in the environment or restoring or expanding the role of nature in the environment. True wilderness is increasingly rare, though wild nature can be found in many locations previously inhabited by humans.

Although industrialization is important for the economic growth and development of a society, it is generally harmful to the environment. Industrial processes can lead to climate change, pollution to air, water and soil, health issues, extinction of species, and so on. International focus on global climate change and its effects on society as well as the environment has resulted in an increased focus on the subject of industrial and other pollutants (Rykowska and Wasiak, 2015). The approach of contaminant management can be separated into two distinct categories: the reduction of the emission of pollutants from the immediate sources, such as engines or factories, and remediation of existing environmental pollution through active clean-up techniques (Hammer, 2014; Kim et al., 2016a). For example, we have the treatment of exhaust flue gases for engines or generators, processing wastewater from industrial plants, carbon sequestering from atmospheric CO_2, and soil remediation from past contamination such as those of the so-called "Superfund sites" within the U.S. Recently, many CAP applications toward the reduction of pollutants, especially in gaseous or aerosol form, involve plasma-assisted or plasma-directed catalysis (Thévenet et al., 2014; Vandenbroucke et al., 2011; Whitehead, 2010; Zadi et al., 2020). As catalysis is approached with more depth in a later chapter in this book, the focus of this chapter will be on applications and general methods rather than the specifics of the reactions themselves. In addition, CAP applications have been developed for renewable power technologies and cleaner-burning fuels in conjunction with pollution and contaminant management across a broad range of categories (Hammer, 2014; Rykowska and Wasiak, 2015). Here, we present CAP technology for three major classifications in the environment: air, water, and soil.

5.1   AIR POLLUTION

Air pollution occurs when harmful or excessive quantities of substances are introduced into the Earth's atmosphere, including organic solvents, respirable particles, sulfur dioxide, nitrogen oxides, etc. These pollutants can cause diseases, allergies, and even death to humans as well as damage the environment by contributing to global phenomena such as climate change, the greenhouse effect, the ozone depletion, and increasing desertification (Birdsall and Wheeler, 1993; Brunekreef and Holgate, 2002; Dockery et al., 1993). Indoor air pollution and poor urban air quality are two of the world's worst toxic pollution problems as listed in the 2008 Blacksmith Institute World's Worst Polluted Places report (Ericson et al., 2008). Air pollution in 2012 caused the deaths of approximately 7 million people worldwide according to the 2014 World Health Organization report (Organization, 2018).

Aerosolized and gaseous contaminants are the most broadly studied subject for CAP pollution treatment, as the nature of CAP generation often includes the active feeding of gas into the reactor. Further, plasma discharge can interact with subject gases three-dimensionally rather than only at a (largely) 2D surface, leading to more efficient processing. For example, removal of various common indoor pollutants from the air and air-mixed gases using corona and DBD reactors found a wide range of efficiencies but the generally productive application of the technologies (Bahri and Haghighat, 2014). The same team found that ozone concentration (one of the primary measures of efficiency) increased with the size of the inner electrode in a DBD reactor and decreased with increases in the gap unless additional input energy was supplied to the reactor, highlighting the complex 3D nature of treatment (Bahri et al., 2016). A similar study comparing the removal of various volatile organic compounds (VOCs) under constant parameters found that removal efficiency via CAP trended with the molecular weight percent of hydrogen in the pollutant molecule (Karatum and Deshusses, 2016). Moreover, Ma et al. indicated factors, namely the physical structure of the reactor and various power supply parameters, that significantly affect removal efficiency (Adnan et al., 2017; Ma et al., 2017; Malik et al., 2011). The complex interactions, both synergistic and inhibitory, between various flue gases (especially O_2, SO_2, NO, H_2O, and water vapor) may explain conflicting results from other studies, as the specific concentrations of each gas likely vary from study to study.

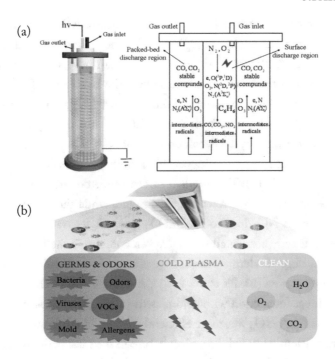

Figure 5.1: CAP for air pollution management. (a) Hybrid surface/packed-bed discharge plasma generator for the degradation of benzene. Reproduced from (Jiang et al., 2013b). (b) Cold plasma employed in air conditioning for purification: filtering harmful/airborne particles, controlling odor, and killing viruses, bacteria, and mold.

Figure 5.1a shows an innovative plasma reactor generating hybrid surface/packed-bed discharge plasmas employed for the degradation of benzene (Jiang et al., 2013b). CAP displayed remarkably better benzene degradation than standard methods, and multiple experiments reported similar results for toluene removal (Feng et al., 2018; Karatum and Deshusses, 2016; Subrahmanyam et al., 2010; Sun et al., 2012b; Wang et al., 2015). The formation of byproducts, such as NO_2, was suppressed, while O_3 was promoted. N_2, as the carrier gas, was found to be effective for benzene degradation because of the fast reaction rate of N_2 $(A^3\Sigma_u^+)$ with benzene, and oxygen species derived from the dissociation of O_2 were found to be significant in the mineralization process. Thus, the addition of O_2 to N_2 allowed for efficient degradation of benzene, and the optimized amount of O_2 was determined to be 3%. A study on the use of CAP to treat diesel exhaust found an inverse correlation between ozone production and particulate matter removal from the exhaust gases, with removal efficiencies over 50% achieved (Babaie et al., 2015). The specific nature of particulate matter created by a diesel engine has been shown to vary with engine load and thus the reaction and byproducts from treatment with cold plasma also change (Gao et al., 2017). Here, higher engine load resulted in increased VOCs released after CAP treatment due to changes in temperature of burn within the engine at higher loads. The creation of and interaction between specific radicals as a result of plasma treatment of VOCs can have synergistic effects when multiple VOCs are processed in the mixture. In one setting, the treatment of a combination of three VOCs in mixture led to a higher conversion percentage for each gas and improved selectivity for CO_2 than separate treatment of the individual gases at the same energy input (Karuppiah et al., 2012b; Kim et al., 2016a). Similarly, the decomposition of NOx was studied in a "wet" cold plasma reactor, with a sodium sulfite solution acting as the humidity source (Cui et al., 2019; Jolibois et al., 2012; Talebizadeh et al., 2014). The problem of excess nitrous oxide being present as a reaction byproduct was tempered by the presence of the liquid, while the specific ions in the solution facilitated the conversion of NO_x ions to N_2 and O_2 gases. A separate study found that removal efficiency of NO_x was directly correlated to the oxygen content of the plasma reactor in addition to determining that surface discharge reactors were more energy-efficient in NO_x conversion than volume discharge reactors (Malik et al., 2011). A comparison of the treatment of NO_x and SO_2 either through direct interaction with cold plasma or through exposure to plasma-generated ozonized air determined that NO_x was more efficiently decomposed through the indirect exposure while SO_2 was more efficiently treated through exposure to the plasma discharge (Obradović et al., 2011). The difference is likely due to the decomposition pathways of NO_x, with additional radicals created when exposed to the plasma directly and, thus, a decrease in overall removal efficiency, while SO_2 is decomposed more by direct activation of the plasma than by ozone exposure. The addition of NH_3 to the mix improved NO_x decomposition in all scenarios, with more of an effect seen for higher initial NO_x concentrations. Another complication for cold plasma treatment of air pollutants is the determination of the efficiency of the system. As pointed out by Kim in an early review, tests employed for determining the concentrations of pollutants after processing need to be carefully selected to prevent artificial results (Kim, 2004). Additionally, the reactions themselves are obviously dependent upon the characteristics of the pollutants and mix gases; for example, removal of NO_x compounds is most generally successful in the presence of increased O_3 creation by the plasma reactor, something that can be more easily achieved with plasma-directed catalysis (Kim et al., 2016a; Obradović et al., 2011).

Figure 5.1b shows CAP applied in an air conditioner for purification: filtering harmful/airborne particles, controlling odor, and killing viruses, bacteria, and mold. The reactive species and ions generated by plasma break down polluted gases to harmless compounds prevalent in the atmosphere such as O_2, N_2, water vapor, and CO_2. For example, plasma can break down formaldehyde into CO_2 and water vapor, thus eliminating the health hazard (Xu et al., 2007). Plasma-generated positive and negative ions attach to airborne particles by their electrical charge, and particles grow larger by drawing nearby particles of the opposite polarity, resulting in their filtration (Park et al., 2011). In addition to particulate gas pollution treatments, CAP has also been investigated as a tool for biological decontamination. A study on the removal of bacterial and fungal aerosols from indoor and outdoor air found that cold plasma treatment for 0.12 s resulted in 98% or greater destruction of the target species (Liang et al., 2012). At 0.06 s exposure, 85% of fungal and 95% of bacterial contaminants were eradicated. CAP-generated ROS/RNS plays an important role in the inactivation of the airborne virus, bacteria, or mold spore (Aboubakr et al., 2016). Additionally, similarly to airborne particles, plasma-generated positive and negative ions also attach to pathogens. Once the ions combine on a pathogen's surface, they eliminate hydrogen from the pathogen, which leads to the airborne virus, bacteria, or mold spore becoming inactive. Thus, the use of CAP in an air conditioner can filter harmful particles from the air within a home, improving air quality and safety.

5.2   WATER POLLUTION

Water is an indispensable requirement of life, and so industrial processes that result in wastewater create an alarming situation requiring immediate attention because of the adverse effects of wastewater pollutants on human and aquatic life. The major group of pollutants in wastewater is organic compounds, which cause severe problems for the environment and human health around the world (Capocelli et al., 2012). These compounds must be treated to satisfy the stringent water quality regulations for aquatic ecosystems. Advanced oxidation processes are innovative tools for such treatment that rely on generating highly reactive species, especially hydroxyl radicals and ozone (Andreozzi et al., 1999; Reddy and Subrahmanyam, 2012; Svarnas et al., 2020; Van Nguyen et al., 2020; Xu et al., 2020). Currently, wastewater treatment methods include anaerobic processing, electrochemical treatment, chlorination, ultrafiltration, radiation treatment, heat treatment, and various combinations of these (Patange et al., 2018). These technologies face major limitations due to increasing costs, residual chemicals in the products, high energy consumption, and the stringent standards of wastewater treatment. Thus, environment management systems and public health authorities must explore innovative methods to reduce the impact of wastewater discharged from organic-intensive industries on the environment. As an advanced oxidation processes, CAP use has increased enormously for the treatment of organic pollutants, mainly due to environmental compatibility and high removal efficiency. Figure 5.2a shows cold plasma generated in a wastewater reactor that can remediate water with a fast treatment rate and environmental compatibility.

Several general factors may affect the efficiency of plasma treatment: the contact area between plasma and water (larger is better); electrode composition and gap distance (smaller is better); energy input to the reactor (larger is better); solution pH (varies); reactor temperature (varies); source mix gas (varies, though oxygen-rich tends to be better); solution conductivity (varying threshold); and target pollutants (Jiang et al., 2014a). Many of these factors are also common to general plasma reactions regardless of medium, though some are particular to the inclusion of a liquid solution in the reaction. Reactor design for water treatment generally creates a thin film of fluid, either by rotating drums or via waterfall, which is then subjected to plasma. Thus, the total volume of treated fluid at any one time is reduced, necessitating an increase in the exposure time for the overall sample in order to reach the necessary degradation state. Increased oxidation rate can be activated by bubbling oxygen gas through the solution and, thus, causing more direct interaction with and oxidation of the target compounds (Magureanu et al., 2011). The most desirable reactive species in plasma is the hydroxyl radical, which reacts with most organic pollutants either by hydrogen abstraction (with saturated aliphatic hydrocarbons and alcohols) or electrophilic addition (with unsaturated hydrocarbons) (Lukes et al., 2012). Another important species generated in the gas-liquid plasma system is hydrogen peroxide, which can decompose most organic pollutants (Locke and Shih, 2011; Malik et al., 2001).

Figure 5.2b elucidates the effects of air CAP byproducts on the degradation of a sample textile pollutan (Chen et al., 2009a). In this study, the gaseous byproducts of a plasma fabric treatment process were applied to water contaminated with methyl violet 5BN to determine their effectiveness in combination with ultraviolet light exposure. The chromophores of azo bonds (–N = N–) responsible for the violet color were broken down when the dye molecules were exposed to the ultraviolet light and ozone generated in air cold plasma. Twenty-five minutes of exposure removed all traces of the primary N-bonded molecules. Further, replacement of air as the cold plasma feed gas with oxygen completely eliminated aromatic byproducts from the secondary reaction (Bruggeman and Leys, 2009). Relatedly, a comparative study of treatment of four sample textile dyes resulted in a significant reduction in toxicity of the final treated water when compared to pre-treatment (Tichonovas et al., 2013). The high initial lethality of Astrazon Brilliant Red was suspected due to the cyanide group present in the dye, which plasma treatment largely neutralized. Figure 5.2c shows ozonation and different advanced oxidation processes such as photocatalytic oxidation, photocatalytic ozonation, with CAP being examined for the degradation of the non-steroidal anti-inflammatory drugs diclofenac and ibuprofen in aqueous solution (Aziz et al., 2017). The mineralization efficiency of each oxidation method was characterized by total organic carbon removal, with the highest removals obtained by photocatalytic ozonation and by cold plasma in an Ar/O_2 atmosphere. In addition, another study compared the degradation of ibuprofen in particular in three configurations: through a Fenton catalytic reaction alone, through a plasma reactor, and through the combination of both (Marković et al., 2015). They found that the Fenton reactor alone had the highest initial efficiency but plateaued quickly at a specific rate, whereas the best overall efficiency was through the combination of Fenton reactor and plasma.

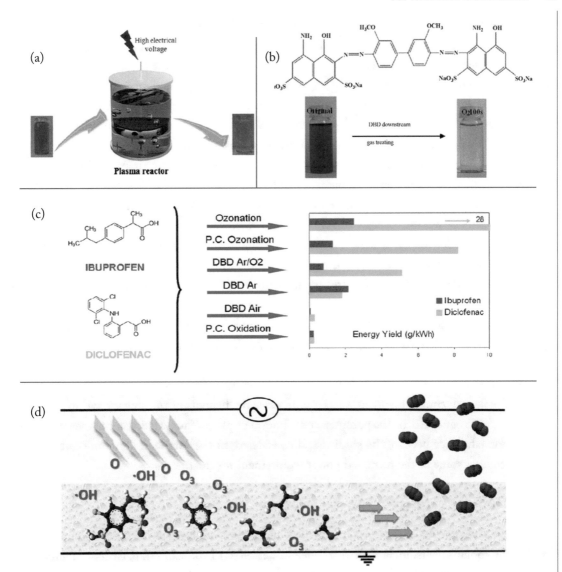

Figure 5.2: CAP for water pollution management. (a) Wastewater remediated with fast treatment rate and environmental compatibility by plasma discharge in a reactor. Reproduced from (Jiang et al., 2014a). (b) The molecular structure of MV-5BN and its color change treated by cold plasma downstream gas. Reproduced from (Chen et al., 2009a). (c) Photocatalytic ozonation, photocatalytic oxidation, and cold plasma examined for the degradation of non-steroidal anti-inflammatory drugs diclofenac and ibuprofen in aqueous solution. The highest total organic carbon removal was obtained by photocatalytic ozonation and by cold plasma in an Ar/O_2 atmosphere. Reproduced from (Aziz et al., 2017). (d) Schematic for complete mineralization of organic pollutants in water treatment by a cold plasma-based advanced oxidation process. Reproduced from (Ceriani et al., 2018).

The removal of pharmaceuticals from wastewater is a subject of increasing concern across the world, with an increasing number of various human-generated compounds being identified in groundwater and other sources (Rykowska and Wasiak, 2015). Figure 5.2d shows the complete mineralization of phenol in water using a plasma reactor, with CO_2 and mineralized pollutants as the primary byproducts, at phenol concentrations in excess of general environmental pollution levels (Ceriani et al., 2018). Ceriani et al. also achieved 95% mineralization of organic pharmaceutical hydrochlorothiazide using the same configuration, implying a broader use of plasma as an effective near-total water pollution treatment strategy. The treatment of various veterinary antibiotics present in synthetic wastewater achieved significant results (Kim et al., 2013). Degradation percentages varied for each of several antibiotics tested, but all experienced exponential increases in removal efficiency as reactor power increased, with 90%+ removal efficiencies generally. Magureanu et al. (2015) conducted a broader test on specific compounds including analgesics, antibiotics, anticonvulsants, antihypertensives, contrast media, lipid regulators, and vasodilators; all of the tested compounds were successfully oxidized via CAP treatment, with the final products being mineralization and highly oxidized organics.

An interesting effect of the cold plasma treatment of water is the creation of plasma-treated water (PTW, as described earlier) (Scholtz et al., 2015). In short, certain stable byproducts of plasma exposure remain in the final solution and allow some of the decontamination effects of plasma exposure to continue passively for moderate durations after exposure. In addition, the treatment of biological contaminants in water also receives the benefits of the shock wave and acoustic effects (Scholtz et al., 2015). The potential contribution of plasma-activated water to overall decontamination efficiency needs to be studied and considered, as the extended duration of effect may help to counter some of the increased power requirement for treatment.

5.3 SOIL POLLUTION

The quick development of industrial production and urbanization results in serious soil pollution, and many contaminants, including heavy metals, petroleum hydrocarbon, persistent organic pollutants, and pharmaceuticals, have been continually introduced into plant sites (Kim et al., 2011; Oonnittan et al., 2009; Yuan et al., 2009). These polluted sites are now faced with commercial and residential utilization due to the increasing economic values of lands, but a large number of toxic contaminants in the soil, especially organic pollutants, poses a great threat to human health (Bai and Zhang, 2017; Bezza and Chirwa, 2017; Thepsithar and Roberts, 2006). Soil contaminated with pollutants is generally treated by conventional physical/chemical remediation, incineration, steam injection, soil vapor extraction, bioremediation, and so on (Gong et al., 2017; Hanna et al., 2005; Sandu et al., 2017; Thuan and Chang, 2012). However, there exist some drawbacks to these processes, such as secondary pollution, long cycle time, low efficiency, and other limitations. Recently, cold plasma as one of the advanced oxidation process technologies has been applied to remediate polluted soil with high degradation efficiency. Several papers on the subject of remediation of pollutions in soils highlight some of the difficulties as well as advantages of the method of cold plasma treatment of soil (Aggelopoulos et al., 2016, 2013; Bali et al., 2017; Li et al., 2014b; Wang et al., 2014a).

Figure 5.3a shows the mechanism by which cold plasma remediates atrazine-polluted soil in a plane-to-grid plasma reactor operating with air at atmospheric pressure (Aggelopoulos et al., 2018). Several degradation pathways of atrazine were identified, with •OH, H_2O_2, and O_3 as the main oxidizing agents. For •OH-dominated processes inducing the degradation of atrazine, the predominant species formed are partially or fully dealkylated atrazine derivatives (Acero et al., 2000; Balci et al., 2009). •OH may attack the carbon atom bearing the chlorine substituent and perform an aromatic nucleophilic substitution reaction thus generating hydroxy-atrazine. In addition to typical •OH degradation pathways, a specific pattern of degradants generated under plasma treatment of atrazine involve singlet oxygen ($^1O_2$) (Aggelopoulos et al., 2018). Singlet oxygen involvement was verified using 9,10-diphenyl anthracene to produce endoperoxide in the Diels-Alder reaction. Atrazine degradation efficiency is an increasing function of applied voltage/discharge frequency and treatment time, which is in agreement with other recently published results (Aggelopoulos et al., 2015a, 2015b; Lou et al., 2012). Residual atrazine in the soil after CAP treatment was well below the safety limit for crop production, aquatic life, and human health.

Figure 5.3b outlines remediation of soil contaminated by fluorene using cold plasma: reactions between active species and fluorene during ex-situ discharge, in-situ discharge, and in-situ discharge and periodic washing (Zhan et al., 2018). Periodic plasma treatment and washing, removing oxide layers generated on the surface of pollutants during the plasma treatment, reduced the discharge time to 45 min, and improved the pollutants degradation efficiency to 99%. The reaction between short-lived active species and fluorene molecules resulted in a series of chemical reactions including substitution and lactonization (Czaplicka and Kaczmarczyk, 2006; Mangun et al., 2001; Qu et al., 2013). Figure 5.3c elucidates plasma remediation of glyphosate-contaminated soil (Wang et al., 2016b). CAP treatment led to changes in the chemical structure of the target pollutant. No new peaks appeared for the plasma-treated soil samples compared with the untreated (as shown in Figure 5.3c I). Inferring the formation of by-products from FTIR spectra is difficult, necessitating that other techniques be used to determine the details of degradation intermediates. Mineralization of glyphosate in the soil after plasma treatment was determined through the formation of NO_3^- and PO_4^{3-} and the removal of total organic carbon (TOC) and chemical oxygen demand (COD). Increasing treatment time beyond 30 min reduced removal efficiencies of TOC and COD (as shown in Figure 5.3c II), which could be attributed to the generation of carboxylic acids in the soil, which are more resistant to mineralization (Wang et al., 2010). As shown in Figure 5.3c III, some degradation intermediates were detected, such as acetic acid, aminomethylphosphonic acid, glycine, formic acid, NO_3^-, and PO_4^{3-}. According to the formation of byproducts, the C-N and C-P bonds of glyphosate molecules in the soil may be attacked by active species, resulting in the breakage of these bonds and the generation of aminomethylphosphonic acid, glycine, formic acid, and acetic acid, accompanied by the release of PO_4^{3-} and NO_3^- (Barrett and McBride, 2005; Chen et al., 2007; Ndjeri et al., 2013). CAP exhibits high potential for remediation of soil pollution because various active species and physical effects could be generated, allowing pollutants in the soil to be removed effectively and rapidly.

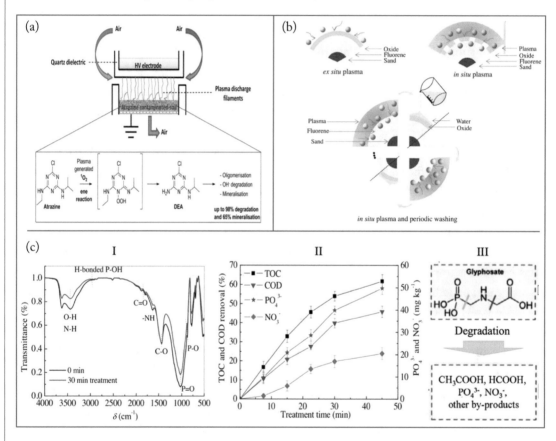

Figure 5.3: CAP for soil pollution management. (a) Cold plasma (DBD) remediation of atrazine-polluted soil in a plane-to-grid plasma reactor operating with air at atmospheric pressure. Reproduced from (Aggelopoulos et al., 2018). (b) Remediation of soil contaminated by fluorene using cold plasma: reactions between active species and fluorene during ex-situ discharge, in-situ discharge, and periodic washing. Reproduced from (Zhan et al., 2018). (c) FTIR spectra of glyphosate-contaminated soils before and after treatment (I), the evolution of TOC, COD, PO43-, and NO3- with treatment time (II), and mechanism of plasma degradation of glyphosate (III). Reproduced from (Wang et al., 2016b).

The remediation of air pollutants has a high potential for advances in CAP technologies. Further research is needed to determine the desired results at a given energy expenditure for targeted pollutants and the range of results needs to be consolidated into models usable at commercial scales. However, even empirically demonstrated exhaust processing via nonthermal plasmas can be of immediate use for high-polluting sources such as power plants and chemical manufacturing. Soil remediation technologies need to significantly improve inefficiency to compete with current treatment options. For example, the investment tradeoff in time versus power for treatment of soil still carries a high burden since driving a plasma discharge into or through soils is significantly more difficult than through fluids. The use of PTW as a potential intermediate for soil treatment seems a likely candidate for more research, as treating bulk volumes of soils, perhaps even in-situ, may be easier via water impregnated with reactive species than relying on gases or direct discharge exposure. Further, PTW could be used in areas without harming the existing plant or animal wildlife and may in fact improve soil conditions generally due to the benefits demonstrated for agricultural applications.

5.4   SUMMARY AND FUTURE DIRECTIONS

The implementation of CAP in various industries has potentially wide-reaching consequences. For example, carbon capture is one of a growing number of concerns for various industries as environmental regulations increase. A proof-of-concept system for not only the capture but also utilization of CO_2 in the industrial exhaust, using plasma catalysis, has been demonstrated by Chen et al. (2020b) and could provide reduced cost in both environmental credits and fuels. The same team also reported on the use of CO_2 via hydrogenation to produce methane using CAP-treated metal-organic frameworks, showing improved selectivity compared to more traditional or thermal catalysts (Chen et al., 2020c). The complex reactions within CAP-driven remediation and decontamination necessitate increased study to ascertain the exact interactions and effects of various parameters on individual system performance. For example, a team developing a CAP-based air purification system designed to clean acetone and methanol determined that increased presence of acetone reduced the effectiveness of methanol degradation (Giardina et al., 2020). Comparison of the decontamination of real diesel engine exhaust between direct and indirect plasma processing showed higher removal rates for NO_x and degradation of CO under indirect methods but reduced removal rates for UHC and CO_2, demonstrating that understanding of the reactions and their products is critical for CAP-related processes (Khani et al., 2020). For a proof-of-concept water treatment plant, conductivity was determined to be the central parameter, with hydroxyl radicals and solvent electrons being the species that most affected the effectiveness of the system (Schönekerl et al., 2020).

CHAPTER 6

Plasma Materials

Low pressure plasmas (LPP) for advanced materials synthesis have played a transformational role in various industries, such as coatings, the semiconductor industry, renewable energy, and liquid crystals displays (Lieberman and Lichtenberg, 2005). In applications of LLP to materials processing, producing or accelerating chemical reactions at the substrate's surface immerses in the plasma can induce the production of new materials, modification, and etching of materials. The materials processing contacted by LPP depends on fluxes of neutrals, including ions, reactive species, electrons, and photons (Oehrlein and Hamaguchi, 2018). Atoms sputtered to form a solid target material by energetic ion bombardment generated by LLP may be deposited on the substrate. LLP exhibits its huge advantages in the deposition of high-quality coatings (e.g., high hardness and adhesion, high density, crystallinity, etc.) over large areas, while CAP has great potential in materials synthesis such as biomaterials and polymers (Dufay et al., 2020; Mattox and Mattox, 2003; Zhu et al., 2020). Applications of CAP in the synthesis or modification of various materials have been the subject of research efforts in the last two decades, through the exact physics underlying the interaction of plasmas with surfaces have yet to be fully determined (Alemán et al., 2018; Chang et al., 2020; Cvelbar et al., 2019; De Smet et al., 2018; Dong et al., 2020; Felix et al., 2017; Kehrer et al., 2020; Mariotti and Sankaran, 2011; Medvecká et al., 2018; Praveen et al., 2019). While the explicit focus on the underlying physics of CAP has decreased in recent years, research into new applications and implementations continues to be conducted (Adamovich et al., 2017). Multiple configurations have been used for materials modification and synthesis. Examples include DBD (Al-Maliki et al., 2018; Astafan et al., 2019; Dimitrakellis and Gogolides, 2018a; Gupta et al., 2015; Kim et al., 2017, 2013; Lang et al., 2018; Mertz et al., 2018; Pandiyaraj et al., 2019; Pavliňák et al., 2018; Shekargoftar et al., 2018; Snoeckx et al., 2015), plasma jet (Abuzairi et al., 2016; Arik et al., 2019; Bartis et al., 2016; Chen et al., 2015a, 2015b; Gerullis et al., 2018; Jurov et al., 2019; Kostov et al., 2014; Lee and Jeong, 2018; Liu et al., 2018; Park et al., 2019, 2018; Roth et al., 2018; Xu et al., 2016a, 2016b; Yáñez-Pacios and Martín-Martínez, 2018), plasma-initiated chemical vapor deposition (Loyer et al., 2018a, 2018b), plasma spraying (Ambardekar et al., 2018; Wallenhorst et al., 2018; Wen et al., 2017; Zhang et al., 2015d), corona (Fabregat et al., 2017; Hawtof et al., 2019), glow discharge (Khlyustova et al., 2019b; Ouyang et al., 2017; Park et al., 2016; Pastor-Pérez et al., 2018; Peng et al., 2017; Yang et al., 2018a), and gliding arc discharge (Tiya-Djowe et al., 2019; Zhang et al., 2015a, 2016). Remote methods of plasma treatment such as remote plasma oxidation have also been reported (Chen et al., 2018a; Luan and Oehrlein, 2018). Applications of CAP in materials processing can be distinguishing between the application of CAP for the modification of materials or material properties created through other means, and the wholesale formation of new materials from raw components. While certain applications combine both methods, the operational concerns for each vary. Further differentiation can be made between the synthesis of bulk material and the creation of nanoparticles, which is significant enough a subset of materials synthesis for a separate discussion.

6.1   SURFACE MODIFICATION

In general, surface modification refers to affecting the optical, electrical, chemical, or structural properties of a material without changing the principal composition. Foest et al. (2005) identified five major categories of surface modification: (1) Etching—removal of measurable mass from the surface of solid material; (2) Functionalization—changes in the surface functional groups of material, including the addition of new functional groups; (3) Cleaning—removal of a measurable mass of contaminants or impurities; (4) Deposition—the creation of nanometer or micrometer films on surfaces; and (5) Radiation—general exposure to plasma and plasma-generated radiation, primarily UV radiation. One general consideration in the application of CAP for surface modification is the size of the affected area. For example, most plasma jet treatments are characterized by a highly localized transformation region, with multiple applications or simultaneous systems needed for larger surfaces. This introduces a limit in the development of industrial-scale applications for plasma jet treatment (Kostov et al., 2014). While arrayed systems can cover larger contact areas, the parameters affecting the interaction of such arrays is still being studied, proposing lower voltage, higher gas flow rate, and specific gases as preferential for arrayed systems (Liu et al., 2018). It has also been suggested that surface irradiation area is dependent upon not only the properties of the plasma jet array itself but on the subject material. For example, surface charge buildup on polymeric subjects may help to disperse the plume and effectively widen the area of applications (Abuzairi et al., 2016). Due to their size difference when compared to traditional materials, modification of nanoparticles poses a field of specific interest for plasmas. As one example, the sintering of silver nanoparticles using glow discharge resulted in similar surface plasmon resonance when compared to thermal methods (Sonawane et al., 2018). Adhesion with other materials has also been subjected to study. AA5052 was treated with CAP and one of three gases (argon, nitrogen, and air) with the resulting surface changes studied under atomic force microscopy and X-ray photoelectron spectroscopy (Wang et al., 2018c). It was found that nitrogen and air treatments resulted in the creation of an aluminum oxide layer rich with other ionic species that increased adhesion, while the argon treatment caused removal of carbon contamination at the surface but did not affect the surface otherwise. The surface treatment of 3D scaffolds and other structures may be of particular interest for plasma research, as plasma discharge may react with interior surfaces not otherwise accessible. On this subject, Daria et al. (2018) conducted a thorough experiment with carbon and ZnO structures exposed to diffused co-planar surface area discharge with varying gas compositions and the introduction of additional species for surface implantation. Their analysis included optical emission spectroscopy study of the plume itself as well as analysis of the resulting surface changes, showing for example that the discharge itself built up 3 mm above the surface of the material.

The act of ablating, vaporizing, or otherwise eroding the surface of a material has multiple uses in industry. While macro-scale surface changes can be achieved quite easily through traditional methods, changing surface morphology at the micro-/nano-scale has been more difficult (Puliyalil and Cvelbar, 2016). Many processes use high-temperature methods or corrosive (and thus generally toxic) chemicals to create such changes, both of which have specific negatives in their implementation (Scholtz et al., 2015). To accurately compare the success of plasma etching on various substances, cold atmospheric plasma was applied to graphene, polymers, proteins, and bacteria (Kuzminova et al., 2017). A sample protein solution showed rapid mass loss initially with a decrease over time due to decreased surface exposure to the plasma; removal rates were still high enough to effectively decontaminate equipment with a few minutes' treatments. SEM images of B. subtilis spores before and after 1 min of CAP treatment are shown in Figure 6.1a. Dramatic changes in the morphology of spores and their size as well as substantial damages to their shells were observed after their exposure to the plasma. CAP-generated ROS and RNS induced the erosion of spores with clearly visible defects in the spore shell (Fumagalli et al., 2012; Park et al., 2003; Vujošević et al., 2007). For bacterial spores, physical removal through etching appears to be separate from the inactivation of spores through radicals produced by the plasma but was still significant in decontamination. There are numerous potential applications for such micro- and nanoscale modifications, for example, creating dopamine-sensing electrodes from both conducting and, more interestingly, electrochemically inert polymers through the application of a corona-discharge plasma in an inert nitrogen atmosphere (Fabregat et al., 2017). In particular, the plasma discharge increased pore size and tortuosity of the surface of the electrodes; these "edges" increase oxidation and reduction processes for molecules in contact with the surface and, thus, increase the effectiveness of the electrodes in dopamine oxidation detection.

The modification of polymers via CAP has multiple potential effects related to the specifics of the treatment. When etching polymers, the rate of etching was linear and appeared related to the oxygen content of the polymer (Puliyalil and Cvelbar, 2016). CAP treatment of polymers and other materials can also result in functionalization changes to hydrophobicity or hydrophilicity, which have ramifications for adhesion, wettability, and other target characteristics (Wolf and Sparavigna, 2010). These changes are generally described by the water-contact angle and are associated with Lewis basicity and acidity (proton accepting/electron-donating and proton donating/electron-accepting respectively) (Good, 1992). Hydrophobicity can therefore be modified through the interaction of the material with high-energy electrons in the plasma discharge or through the modification of existing functional groups on the surface of the material (Bružauskaitė et al., 2016;

Novák et al., 2018; Tiya-Djowe et al., 2019). Chen et al. applied a plasma jet to a nylon mesh to effect superhydrophilicity and superoleophobicity (as shown in Figure 6.1b), with the resulting mesh capable of 97.5% separation of an oil/water mixture (Chen et al., 2016b). Figure 6.1b shows: (I) time series of a water droplet contacting with the plasma functionalized nylon mesh in the air; (II) dichloromethane droplet making and losing contact with a plasma functionalized nylon mesh in water; and (III) time-lapsed snapshots of dichloromethane and hexadecane droplets rolling on a tilted plasma functionalized nylon mesh in water. The surface of the plasma-treated mesh became rougher, with submicron scale pits and flocculent particles. The superhydrophilicity and micro-/nanostructures were crucial for the functionalized mesh to absorb and lock water to form a repellent barrier to oils, thus creating underwater superoleophobicity (Chen et al., 2016a; Fritz and Owen, 1995; Hossain et al., 2019; Nguyen et al., 2014). A similar result was obtained using aerosol-assisted plasma treatment to create nanocomposite membranes with tunable permeability properties for use in structural separation of water and organic solvents (Chen et al., 2018b).

Figure 6.2c shows AFM images of polylactic acid 3D printed scaffolds with and without CAP treatment at different times, which confirms that CAP treatment increased nanoscale surface roughness and decreased water contact angle from $70 \pm 2°$ to $24 \pm 2°(58)$ (Wang et al., 2016a). Minor decreases in contact angle below these values were temporary and potentially due to polymer/electron interaction (Kusano, 2014). Such a polymer/electron interaction would be consistent with other findings, where increases in carbon nanotube arrays subjected to an N_2-fed plasma jet were related to both structural changes in the arrays as well as electron doping (Lee and Jeong, 2018). Following treatment, the water contact angle shifted slightly toward the original value with storage time but remained largely stable. In another study, changes to water contact angle were determined to be maximized by a combination of increased oxygen intensity and decreased application distance between the plasma source and a PET sample (Scally et al., 2017). Sachs et al. further tested the application of plasma jet with a fluid reactor bed in modifying the surface wettability of polymer powders, determining not only that optimal treatment times exist but similarly reporting that the use of various gases in the process changes the nature of the reactions (Sachs et al., 2018). Both studies found that an oxygen-containing feed gas enhanced hydrophilic results, attributed primarily to the creation of hydroxyl functional groups on the surface.

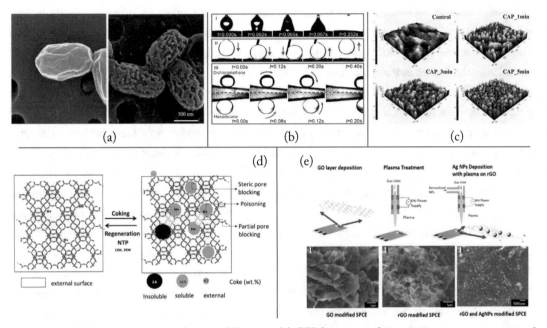

Figure 6.1: CAP for material surface modification. (a) SEM images of *B. subtilis* spores prior to and after 1 min of cold plasma treatment. Reproduced from (Kuzminova et al., 2017). (b) (I) Time series of a water droplet contact with a plasma-functionalized nylon mesh in air. (II) Dichloromethane droplet making and losing contact with a plasma functionalized nylon mesh in water. (III) Time-lapsed snapshots of dichloromethane and hexadecane droplets rolling on a tilted plasma-functionalized nylon mesh in water. Reproduced from (Chen et al., 2016b). (c) AFM images of polylactic acid 3D-printed scaffolds with and without CAP treatment at different times. Reproduced from (Wang et al., 2016a). (d) Complete regeneration of zeolite achieved at 293 K using cold plasma with low energy consumption in a fixed-bed dielectric barrier reactor, a geometry suitable for industrial scaling. Reproduced from (Astafan et al., 2019). (e) Schematic illustration of reduced graphene oxide (rGO) sensor fabrication process: GO deposition, plasma treatment, and Ag nanoparticles deposition on rGO layer, and SEM images of morphological changes during each step of the sensor fabrication for cortisol detection. Reproduced from (Sonawane et al., 2019).

The use of CAP treatment in cleaning and sanitization is also a subject of interest, especially with delicate materials, as often such processes involve harsh thermal or chemical treatments that can damage the material (Kramer et al., 2018; Roth et al., 2018). Multiple laboratory applications of CAP in decontamination and sterilization in the food industry for a wide variety of products (Dimitrakellis and Gogolides, 2018a). Separate implementation of the cleaning of a bulk material has been introduced by the recovery of zeolite to initial condition for use in commercial hydrocarbon conversion (Astafan et al., 2019). Figure 6.1d shows that complete regeneration of zeolite can be achieved at 293 K by using cold plasma with a low energy consumption in a fixed-bed dielectric barrier reactor, which a geometry suitable for industrial scaling. Zeolite functions as a catalyst for propene transformation, and the cavities within the zeolite structure facilitate the transformation but become clogged through the process. It was found that plasma-generated oxygen radicals diffused through the material and reacted with the lighter detritus molecules 36 times more easily than with the principle compounds in zeolite itself (Moljord et al., 1995). In contrast, the creation of surface films using nonthermal plasmas to replace chemical or other harsh methods is another area with a large amount of investigation. While multiple methods have been demonstrated in the lab, specific studies now focus on industrial-level implementation, such as a recent study on the fabrication of graphene oxide films for a variety of uses (Alotaibi et al., 2018). Figure 6.1e shows the schematic illustration of reduced graphene oxide (rGO) sensor fabrication process: untreated GO exhibited 2D planar, packed layer structure and aggregated sheets closely associated to form a continuous structure, whereas the plasma modified GO possessed a wrinkled, folded structure. The reason for the comparatively continuous, soft morphology of GO layers might be because of the presence of hydroxyl or carboxyl groups (Gurunathan et al., 2015). The observed change in surface roughness of the rGO was due to the plasma bombardment (Li et al., 2014a). A related study demonstrated the formation of reduced GO-silver nanoparticle composites in surface layers for biomedical sensors (Sonawane et al., 2019). A further study on inkjet-printed GO fibers determined that plasma treatment had limited negative effects on the treated fibers, thus suggesting the process to be safe for even more delicate samples (Homola et al., 2018). The deposition of thin layers of nebulized or gaseous substances is a common cause of CAP implementation, such as coating common hydrophilic membranes with a micrometer-scale hydrophobic plasma-deposited layer for use in desalinization (Eykens et al., 2018; Foest et al., 2005). In addition, another method of modification of polymers with atmospheric plasma is the plasma-initiated chemical vapor deposition (Loyer et al., 2018a, 2018b).

Surface modification through CAP-generated radiation treatment has been primarily studied in the areas of medicine and food processing. Given the production of both radiation and reactive species in the plume, isolation of the effects of radiation on materials must be performed carefully (Schlüter et al., 2013). For precision tools utilizing miniaturized plasma jets, the intense UV and VUV radiation play an important role in surface modification (Foest et al., 2007). The effects of plasma-generated UV and VUV radiation on microbial species have been studied, with most reports indicating successful inactivation of harmful microbes and spores (Brandenburg et al., 2009). Electrostatic disruption of cellular activity due to interaction with the high-energy electrons in the discharge was determined to play a part in cellular deactivation alongside UV exposure and interaction with radicals (Liao et al., 2017). Treatment of seeds and seedlings with CAP explicitly relies on UV and VUV radiation from the plasma discharge (Bußler et al., 2015). Here, flavonoid content was affected by both dose and timing of the exposure, with a short but high dosage of UV-C affecting specific pathways. The authors suggest that interaction between the UV effects on the seeds and the ROS effects generate different results depending on the development stage of the plant.

6.2   MATERIAL SYNTHESIS

Plasma technology, a low-energy consumption process with limited chemical wastes, has been regarded as a highly efficient tool for material synthesis (Dou et al., 2018; Neyts et al., 2015; Šimek et al., 2019). Figure 6.2 shows CAP technology used for material synthesis. The synthesis of gases via CAP has been widely studied. For example, as a potential replacement for fossil fuels, Peng et al. (2018a) achieved a significant efficiency in ammonia production using plasma technology with a generation rate of 20.5g/kwh. In another example of a catalyst-based system, ammonia synthesis over ruthenium with a magnesium promoter was conducted via atmospheric plasma, with the addition of O_2 gas sufficient to promote regeneration of the catalyst (Kim et al., 2017). Synthesis of ammonia from nitrogen and water without catalysts has been achieved with a reported efficiency nearing 100% in a process combining atmospheric plasma with electrolytic methods (Figure 6.2a); the process required significantly higher energy than other reactions (Hawtof et al., 2019). The action of important species such as vibrationally excited N_2 (contained in the plasma) and solvated electrons [e–(aq)], (contained in the water) and their reaction products (such as hydrogen radicals (•H)) leads to NH_3 formation (Mehta et al., 2018; Van der Ham et al., 2014; Witzke et al., 2012). The overall reactions for N_2 reduction to NH_3 and H_2 evolution (under acidic conditions) at the cathode are shown in Figure 6.2a. Methane has also been studied as a candidate for plasma production, with a detailed analysis of plasma-based dry reforming of methane conducted with the goal of maximizing efficiency: a peak performance of 84% energy conversion and 8.5% energy efficiency was reached (Snoeckx et al., 2015).

   Figure 6.2b shows CAP synthesis of metal and alloy nanoparticles in-situ in polyvinyl alcohol (PVA) hydrogel with potential multifunctionality for a range of biomedical applications (Nolan et al., 2018). PVA serves as a capping agent preventing nanoparticle agglomeration and impedes further interaction between the nanoparticles' nuclei with the reactive species affecting nanoparticle crystal growth kinetics as well as the resulting nanoparticle size and morphology (Kyrychenko et al., 2017; Liu et al., 2010; Nolan et al., 2018; Sotiriou et al., 2012). All composite samples containing Ag nanoparticles exhibit anti-bacterial behavior against both *E. coli* and S. *aureus*, because Ag nanoparticles increase surface area and generate reactive oxygen species (Le Pape et al., 2004; Salem et al., 2015). Figure 6.2c details how thin hydrophobic surface-coating layers are deposited by plasma treatment, controlling the thickness and morphology pattern of the layers to allow water and ion molecules to pass through the coating (Park et al., 2016; Tendero et al., 2006). Plasma-generated reactive species break c-C_4F_8 into various species of fluorocarbon monomers. Two different degrees of functionalization (n = 0.4 or 0.6) were fabricated in sulfonated poly(arylene ether sulfone) in the protonated form (BPSH) or sodium form (CBPS) and an aminated poly(arylene ether sulfone) (ABPS), respectively. XESPSN is an end-group cross-linked sulfonated random copolymer with a degree of sulfonation of 0.6. "Multiblock" refers to a multiblock BPSH–BPS copolymer, which

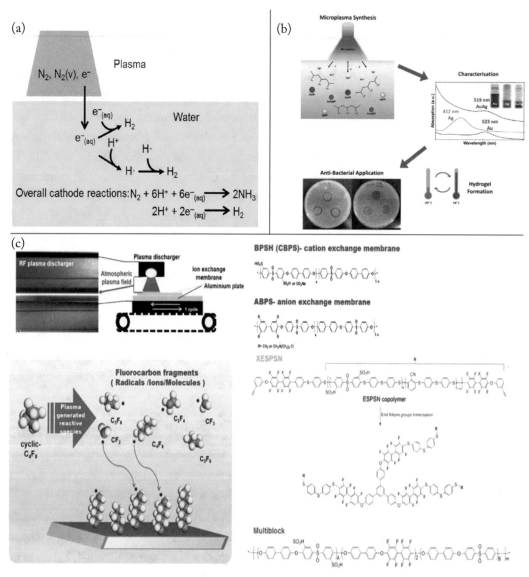

Figure 6.2: CAP for material synthesis. (a) Catalyst-free, electrolytic NH3 production from N2 and water using a plasma electrolytic system. Reproduced from (Hawtof et al., 2019). (b) CAP synthesis deployed for the synthesis of metal and alloy particles in-situ in a polyvinyl alcohol hydrogel with potential multifunctionality for a range of biomedical applications. Reproduced from (Nolan et al., 2018). (c) The purple-lit atmospheric plasma of c-C4F8 and He for polymerization and fluoro hydrocarbon coating layer. Reproduced from (Park et al., 2016).

consists of 10 kg mol^{-1} BPSH and 5 kg mol^{-1} BPS with a spacer linkage. In proton-exchange membrane fuel cells, water retention in the membrane is crucial for efficient transport of hydrated ions (Moghaddam et al., 2010; Service, 2004; Steele Brian and Heinzel, 2001). Plasma polymerization formed a hydrophobic coating layer on the nanometer scale, which helped the membrane remain self-humidified with water generated by electrochemical reactions.

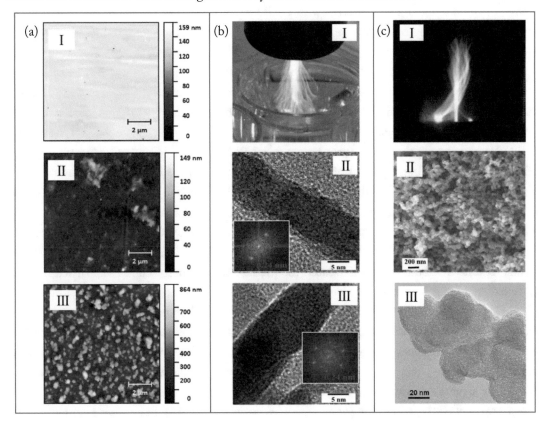

Figure 6.3: **CAP for nanomaterials synthesis.** (a) AFM images of cold plasma jet-assisted impregnation of gold nanoparticles into polyvinyl chloride polymer: untreated sample (I), Au colloid drop-coating (II), and plasma jet treatment (III). Reproduced from (Jurov et al., 2019). (b) Synthesis of copper-based nanostructures in liquid environments using cold plasma jet: plasma jet impinging onto a liquid surface (I), TEM images of nanostructures obtained from the plasma treatment of SOL1 (II) and SOL2 (III) inserting the corresponding fast Fourier transform analysis, respectively. Reproduced from (Liguori et al., 2018). (c) CAP for the gas-phase synthesis of carbon nanoparticles: cold plasma jet operating with argon (I), SEM (II), and TEM (III) micrographs of a furnace-type carbon black nanomaterial. Reproduced from (Moreno-Couranjou et al., 2009).

Nanomaterials have become one of the most active research fields for a couple of reasons: (1) to fabricate new materials on an even finer scale to continue decreasing the cost and increasing the speed of information transmission and storage; and (2) to display novel and often enhanced properties compared to traditional materials, which opens up possibilities for new technological applications (Buzea et al., 2007; Colvin, 2003; Xu et al., 2013). Recently, more publications discussing the production of nanomaterials with CAP have covered a wide range of materials, such as nanoparticles impregnated in polymers, metal nano-structures/composites, and graphene/carbon nanotubes (Bond, 1991; Di et al., 2018; Liguori et al., 2018; Merche et al., 2018; Ni et al., 2018; Nolan et al., 2018; Sun et al., 2016). Figure 6.3a reveals CAP applied for designing polymer/nanoparticle composite materials. Au nanoparticles were drop-coated onto a polyvinyl chloride (PVC) surface with uneven distribution, followed by plasma treatment to drive the formation of Au nanoparticles into an impregnated surface that was mechanically stable and resistive (Jurov et al., 2019). After treatment, the PVC surface becomes rich in N-C = O, C-O, and C = O groups, which led to a cross-linking impregnation of Au nanoparticles to form the nanostructure surface (Phan et al., 2017b; Suganya et al., 2016). Figure 6.3b explains CAP-driven synthesis of a copper-based nanostructure in liquid environments. CuO crystalline structures with limited impurities (e.g., Cu and $Cu(OH)_2$ phases) were produced from solutions of NaCl and NaOH, while a solution of only NaCl produced mainly CuO and $CuCl_2$ structures (Liguori et al., 2018; Velusamy et al., 2017). The plasma treatment of both solutions induced the production of nanorods and nanoparticles. Other oxide nanomaterials have also been produced using gliding arc atmospheric plasma in humid air (Tiya-Djowe et al., 2019). For example, the iron oxides formed hollow "sea urchin-like" structures similar to the "nanocorals." Figure 6.3c investigates CAP for the gas-phase synthesis of carbon nanoparticles. Homogeneous products made of aggregates of nanoparticles were little ramified and linear (Moreno-Couranjou et al., 2009). There were two possible mechanisms: growth by steps in a highly saturated environment and the liquid hydrocarbon droplets theory (Dey et al., 2016; Levchenko et al., 2016; Monthioux et al., 2006; Weissker et al., 2010). In another example, plasma treatment has been tested on biochar for the creation of capacitive carbon for supercapacitor use (Gupta et al., 2015). Here, the standard multi-hour treatment at temperatures in excess of 900°C was replaced by a 5-min treatment with an oxygen plasma jet at atmospheric pressure with activation in excess of the high-temperature method: 171.4 F/g in comparison to 99.5 F/g using the standard treatment. Specific challenges have been presented in working with atmospheric-pressure cold plasmas for nanocrystal regeneration, primarily related to maintaining the large disparity between electron and ion temperatures and gas breakdown at higher pressures (Kortshagen et al., 2016). The smaller volume and continuous gas flow through CAP jet systems allows for additional study and control over nanocrystal formation (Kramer et al., 2015).

6.3   SUMMARY AND FUTURE DIRECTIONS

The intricate link between chemistry and plasma physics is especially prevalent in materials interactions (Weltmann et al., 2019). To most fully explore such uses as surface modification and polymer creation, plasma physicists will have to work closely with chemists and material science experts to determine the mechanisms of reactions and how the specific conditions with the reactor or discharge plume affect the chemical or physical changes that occur. Even a concept as widely studied as changes in water contact angle due to plasma exposure will need far more in-depth analysis to determine exactly why hydrophobicity increases in some situations and decreases in others (Dimitrakellis and Gogolides, 2018b; Huang et al., 2010). Surface modifications can still be done reliably and effectively with CAP limited to "isotropic" processes due to lack of ion bombardment. As a comparison, samples immersed in LLP are subjected to energetic electrons, high-energy ion bombardment, UV light, and reactive species. LLP cleans the sample surface with sputtering while CAP does so without sputtering. Empirical analysis has yielded impressive results but scaling to industrial-level applications will require better understanding and control of the reactions.

The potentials for material synthesis by CAPs are in their interaction with liquid or gels for the generation of nanoparticles, where low-pressure plasmas (LPP) cannot be applied (Wang et al., 2016a). Compared to LLP, CAP only treats samples by reactive gas beam, and leaves the rest of the workpiece untouched. CAP nanomaterial synthesis has vast potential for commercial and industrial use but requires significant research to fully characterize and implement (Dou et al., 2018). Additionally, simplified and more efficient creation of carbon nanoparticles such as nanotubes and fullerenes would have tremendous benefits for many branches of materials science, especially involving composites or energy storage. While precision construction using CAP is likely not feasible, combining the effects of plasma discharges on polymers with various chemical changes due to reactive species may facilitate the deliberate construction of complex micro- or nanostructures with wide ranges of attractive properties (Chen et al., 2016b; Zaitsev et al., 2019). Thus, more research on specifically which characteristics of the discharge result in which properties and the degrees to which those properties can be tuned are needed.

CHAPTER 7

Plasma Catalysis

Catalysis is a mature and important research field that was first reported in 1823 by Johann Wolfgang Döbereiner (Kauffman, 1999). It states that "finely divided platinum powder causes hydrogen gas to react with oxygen gas by mere contact to water whereby the platinum itself is not altered," a process that in 1835 was termed "catalysis" by Jöns Jacob Berzelius. Today, some 80% of all industrial processes are related to catalysis, which suggests a value for catalytic products of approximate $20 trillion per year around the world (Neyts et al., 2015). As a broad concept, catalysis is the acceleration or facilitation of a reaction through the introduction of a substance that is not consumed in the reaction itself (the catalyst). While there are many specific chemical mechanisms by which catalysis takes place, the general result is a reduction in the energy required to initiate the reaction. As energy requirements can often be translated linearly into financial costs, the search of new or better catalysts for industrial processes is a constant effort. Further, catalysts can often reduce harmful or wasteful byproducts of reactions and thus provide additional benefits in a society concerned about global climate change. According to the European Roadmap on Science and Technology of Catalysis, improvements to chemical and petroleum processes using new and improved catalysts could reduce energy usage for production by an estimated 20–40% (Perathoner et al., 2017). Due to reactivity, low-temperature operation, and far-from-equilibrium state, CAP can be used for many complex applications, including catalysis. Thus, plasma catalysis is an emerging research field, which including physics, chemistry, catalysis, nanotechnology, and others.

The use of CAP in conjunction with catalysis can be grouped into two primary forms: plasma modification of catalysts and plasma-catalytic processes (Jang et al., 2004). The first and most straightforward of these is the plasma-based creation or treatment of catalysts for use in separate reactions through generation of nanoparticle catalysts, impregnating nanoparticles onto supports and treatment of the catalysts and supports themselves in ways that augment the catalyzation process (Liu et al., 2002). While most of the methods are similar to other surface treatment and material synthesis procedures, the specifics of the catalytic reactions have implications for its plasma treatment. CAP can also be utilized in catalysis itself in two methods: direct application of plasma to both the subject, usually in gaseous or liquid form, and catalyst in plasma-driven catalysis (PDC), and pre-exposure of the subject to the plasma prior to being introduced to the catalyst in plasma-assisted catalysis (PAC) (Feng et al., 2018; Roland et al., 2005). With PDC, the plasma interacts with the subject directly in the presence of the catalyst. In reactor-generated discharges such as DBD, the reactor itself is generally "packed" with the catalyst, and the subject is cycled through the reactor; Wang et al. utilized this configuration with manganese-copper catalysts to oxidize NO_x

pollutants in gaseous form (Wang et al., 2018b). Alternatively, a typical example of PAC is provided by Chen et al. (2016c) in their work on CO_2 decomposition. Here, a pulsed microwave plasma generator was used to activate gaseous CO_2 immediately prior to its contact with the specific catalyst placed approximately 3 cm below the plasma plume. While many plasma-catalysis systems utilize DBD systems instead of pulsed microwave, the configuration is generally similar. Finally, the use of plasma itself as a stand-alone "catalyst" for some reactions has been studied. While not explicitly catalysis in the classic sense, use of plasma as an alternative for a traditional catalyst has been tested in several situations. For example, multiple studies have examined the use of several distinct cold plasma reactors in hydrocarbon fuel reforming as an alternative to catalyst-based methods currently being analyzed for use in automobiles with encouraging results (Paulmier and Fulcheri, 2005; Petitpas et al., 2007). Similarly, CAP was used to degrade pharmaceutical compounds in water with explicit details on the pathways for various compounds analyzed (Magureanu et al., 2015). Decompositions of volatile organic compounds (VOCs) have also been demonstrated with the application of plasma alone, though with lower efficiency than in conjunction with catalysts as discussed later (Abou Saoud et al., 2020; Karatum and Deshusses, 2016).

7.1   PLASMA TREATMENT OF CATALYSTS

Plasma treatment prior to use has been found to increase the efficiencies and beneficial properties of myriad catalysts across a range of applications. Exposure to discharge and the radicals formed therein, especially within high-oxygen environments, changes the physical characteristics of both catalysts themselves and their support structures in ways that broadly improve the reaction, largely through the increased surface area, increased activation of reactive metals, and changes to the crystalline structure of the catalyst/support material. For example, plasma treatment of palladium catalysts on a titanium oxide support with selected gases was carried out to determine the relative effects of each gas treatment on the catalyst in comparison to untreated samples when used in acetylene hydrogenation (Li and Jang, 2011). The resulting catalysts all performed better than untreated samples, with different gas treatments resulting in slightly different temperature behavior profiles and efficiency. The authors attributed these results to changed Pd activation and modified catalyst/support crystalline structure. Modification of MnO_x catalysts improved low-temperature decomposition of NO significantly, from 5% up to 65% at 50°C (Tang et al., 2012). This increase was attributed to increased pore size, wider pore variability, and increased catalyst dispersion. It was also found that the plasma treatment deposited additional functional groups on the catalysts, further increasing their catalytic activity. Rahemi et al. (2013) treated Ni/Al_2O_3-ZrO_2 catalysts for use in methane reforming and examined the resulting changes via X-ray diffraction, field emission scanning electron microscopy, and other methods. They determined that a 15-min glow discharge treatment with Ar gas resulted in reducing nanoparticle size and increasing surface area. These characteristics led to lower reaction temperatures required for methane reforming and increased efficiency for the reforming process itself.

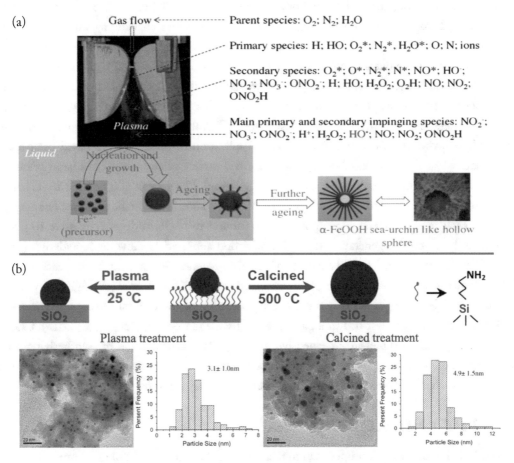

Figure 7.1: CAP for catalyst treatment. (a) Creation of α-FeOOH sea-urchin like nanostructures by plasma: the hydroxyl radical created by plasma is one of the main impinging species that can react with aqueous species at the plasma-liquid interface. Reproduced from (Tiya-Djowe et al., 2015). (b) Particle size changes and TEM images of dried Au/SiO$_2$ particles after O$_2$ plasma treatment and calcination, respectively. Reproduced from (Liu et al., 2012).

The inherent nature of plasma synthesis allows for the creation of novel structures with high surface area, a factor that is especially important for many catalysts. Figure 7.1a conveys the use of a plasma reactor to create "sea urchin-like" goethite structures for the decomposition of organic pollutants in wastewater (Tiya-Djowe et al., 2015). During post-plasma aging, the structures became hollow and porous, increasing the surface area of the catalyst and thus improving their use for catalysis. In general, goethite particles synthesized by plasma-generated ROS/RNS consist of sea-urchin like hollow spheres (He et al., 2002; Pinto et al., 2012). The catalytic activity of such goethite in the Fenton degradation of Orange II (organic dye) can be used as a heterogeneous catalyst for the effective removal of organic pollutants from wastewater. The initial creation of the nanoparticles themselves and impregnation on the support is one of the more challenging aspects of catalyst creation. Figure 7.1b presents O_2-fed RF plasma discharge generating significant Au nanoparticles of small diameter while burning off excess organic material (Liu et al., 2012). This process resulted in a catalyst with higher efficiencies at lower temperatures than its calcinated equivalent, and the authors speculated that the process may be extensible to other metals. Changes to the shape of metal particles or metal-support interaction as a result of CAP treatment could result in more active sites for acetylene hydrogenation (Jang et al., 2008). Application of plasma to Au-based catalysts for CO oxidation was found to result in increased hydroxyl surface groups as well as a large amount of low-coordination number Au species (Zhang et al., 2015c). Plasma can also be used to recharge a catalyst and restore its capability, though perhaps with limitations for specific catalysts. For example, Ag-based catalysts for toluene decomposition were found to begin to degenerate after 10 plasma recharge cycles whereas regeneration through calcination appeared to show no degradation (Wang et al., 2015). On the other hand, H_2-fed plasma catalyst regeneration was found to greatly improve NOx storage capacity for several non-platinum-based catalysts (Zhang et al., 2015f). One potential advantage of PDC configurations is the active regeneration of the catalyst during operation, such as with an AgMnOx catalyst bed in a back-corona discharge reaction (Feng et al., 2013). Chen et al. (2019) proposed a simple, fast, inexpensive, and highly efficient method for synthesizing nitrogen-doped titanium dioxide (N-TiO_2) as an enhanced visible light photocatalyst. CAP can be used as a green and cheap efficient route for the synthesis of porous metal oxide nanostructures.

7.2   PDC AND PAC REACTIONS

The combination of effects from plasma interaction with the subject and potential interaction with the catalyst, either directly or through longer-lived radicals and other byproducts of the plasma-subject reaction, introduces a significant number of variables to PDC reactions. Further, specifics of the catalytic reactions themselves create new challenges in predicting the outcomes of plasma-catalysis reactions that can vary between PDC and PAC. Further complicating the subject is the interaction of the plasma with the catalyst in a broader sense, as PDC configurations are subjecting the catalyst itself to plasma treatment. Thus, they likely change the properties of the catalyst as shown previously. The introduction of the catalyst into the reactor as in PDC configurations requires increased attention to specific details. Figure 7.2a describes ammonia synthesis on wool-like Au, Pt, Pd, Ag, or Cu electrode catalysts in CAP of N_2 and H_2. The activity change is mainly due to migration of metals from the electrode to the inner wall of a silica reactor or increases in surface areas of metal catalysts, whose order at each initial experiment is Au > Pt > Pd > Ag > Cu > Fe > Mo > Ni > W > Ti >

Al (Iwamoto et al., 2017). Plasma processes using various electrodes showed that the efficiency of electrodes for plasma activation of nitrogen molecules was almost independent of the metals (Clark et al., 2005; Gómez-Ramírez et al., 2015). The formation/decomposition rate constants of ammonia were finally determined on Au and Cu, which were different from those for the conventional Haber–Bosch process (Soloveichik, 2019; Vojvodic et al., 2014). Plasma-catalyst interactions work both ways, with plasma affecting the physical properties of the catalyst, and the catalyst changing the nature of the plasma. This latter interaction primarily involves changing the electric field either due to dimensional changes in the reactor or to surface properties of the catalyst itself (Neyts et al., 2015). For example, even the simple physical placement of the catalyst within the DBD reactor can affect the strength of the electric field within the reactor by decreasing the effective gap, thus introducing potential changes to reactions in the gap (as shown in Figure 7.2b) (Wang et al., 2017a). Some evidence suggests in-reactor changes to catalysts in PDC mirror those for direct plasma treatment of the catalyst. The mobility and concentration of O can be improved due to the interaction between Ce and Mn, and a high concentration of toluene is degraded in the plasma-catalytic system with CeO_2-MnO_x catalyst (Liu et al., 2013; Wang et al., 2014b; Zhou et al., 2014). One study found increased oxidation of the metal reactants in the catalyst as well as a decrease to surface area and increase of pore size due to plasma/catalyst interaction (Wang et al., 2017a). These changes are similar to those explicitly induced by the plasma treatment of a catalyst. It increases inefficiency of a PDC process, which must be carefully tested to isolate the treatment of the catalyst itself from other influences. In a separate study, Au-based catalysts for water-gas shift reactions were found to increase gold particle size and subsequently decrease surface area, reducing the performance of the catalys (Stere et al., 2017). These changes may reflect exposure times to the CAP discharge that exceeds standard catalyst treatment times, though more investigation is warranted. Such contradictory results make evident the need to carefully test each aspect of the reaction for individual contribution to the outcome. Separately, the effect of plasma on the catalyst/subject interaction is an area where the synergies in plasma catalysis can be successfully exploited. As described by Neyts et al. (2015), the interaction between the catalyst and species that have been vibrationally excited by the discharge shows that both the forward reaction barrier and activation energy of the catalyzed reaction may be decreased, increasing forward reaction rate without affecting the reverse rate as significantly.

Methane conversion in a DBD reactor also benefits from vibrationally-excited subjects, improving conversion efficiency (Nozaki and Okazaki, 2013). For the reverse reaction, methanation of CO_2 in a PDC configuration showed efficiencies comparable to catalyst-only reactions at far higher temperatures, indicating direct plasma activation of the subject substituted for traditional thermal activation (Nizio et al., 2016). Figure 7.2c indicates plasma synthesis of methanol through plasma-catalytic hydrogenation of CO_2. Compared to the plasma hydrogenation of CO_2 without a catalyst, the combination of the plasma with Cu/γ-Al_2O_3 or Pt/γ-Al_2O_3 catalyst significantly enhanced the CO_2 conversion and methanol yield (Behrens et al., 2012; Wang et al., 2018a; Yang et al., 2015). The synergetic effect of the plasma and catalysts enables selective catalytic CO_2 hydrogenation to methanol to occur at atmospheric pressure with room temperature, providing a new alternative approach that lowers the kinetic barrier and energy cost of catalytic CO_2 hydrogenation for methanol synthesis instead of using traditional approaches (Stere et al., 2015; Whitehead, 2016). Figure 7.2d shows plausible reaction mechanisms of plasma-catalytic removal of formaldehyde over Cu–Ce catalysts in a plasma discharge reactor. The combination of plasma with the Cu–Ce binary oxide catalysts significantly enhances the reaction performance (Zhu et al., 2015). Cu^{2+} sites on the catalyst surface can be reduced to Cu^+ and re-oxidized to Cu^{2+} by the adsorbed oxygen or the lattice oxygen from CeO_2, which suggests Cu^{4+} acts as an oxygen source for the oxidation of HCHO (He et al., 2013). The consumed oxygen species on the catalysts can be replenished by capturing oxygen molecules and reactive oxygen species generated in the plasma (Blin-Simiand et al., 2005; Menon et al., 2012; Nozaki et al., 2007).

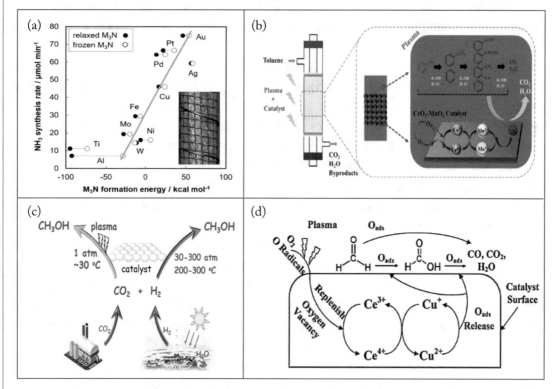

Figure 7.2: CAP for PDC and PAC reactions. (a) Ammonia synthesis on Wool-Like Au, Pt, Pd, Ag, or Cu electrode catalysts in cold atmospheric-pressure plasma of N_2 and H_2. Reproduced from (Iwamoto et al., 2017). (b) Plasma-catalytic removal of toluene over CeO_2-MnO_x catalysts in an atmospheric dielectric barrier discharge. Reproduced from (Wang et al., 2017a). (c) Atmospheric pressure and room temperature synthesis of Methanol through plasma-catalytic hydrogenation of CO_2. Reproduced from (Wang et al., 2018a). (d) Plasma-catalytic removal of formaldehyde over Cu–Ce catalysts in a dielectric barrier discharge reactor. Reproduced from (Zhu et al., 2015).

Additionally, photons generated within the plasma, as well as increased surface collisions, can drive the additional reaction and improve catalytic performance (Van Durme et al., 2008). For example, a TiO_2 photocatalyst embedded in a glass bead dielectric material for a DBD reactor was found to increase catalysis of NO and SO_2 when compared to plasma treatment without the catalyst (Nasonova et al., 2010). Similarly, hydrocarbon production in a PDC configuration over nickel-based support showed marked improvement over catalyst-only and plasma-only reactions (Blanquet et al., 2019). Conversion of CH_4 to CO_2 via CAP and palladium-based catalytic reactions in a PDC configuration was found by Lee et al. to have peak efficiency with a 2%-weight catalyst on aluminum oxide supports preferentially over other supports (Lee et al., 2015). This efficiency was isolated to a specific plasma catalyst reaction by comparison to the same materials in a PAC configuration, which did not show the synergistic results. The authors suggest that the specific short-lived radicals created in by the CAP enhanced the functioning of the catalyst in the presence of the Al_2O_3 support but would dissipate prior to reaching a downstream catalyst in a PAC system.

Placing the catalyst downstream from the plasma-subject interaction, as in PAC, has different implications from the PDC configuration. In some respects, the interaction is similar to the remote plasma oxidation of polymers discussed previously: long-lived reactive species generated within the plasma discharge can be carried by the subject gas to the catalyst chamber and affect the reaction (Luan and Oehrlein, 2018). However, simply the separation of the plasma-subject interaction from the catalytic reaction may be beneficial in certain cases. Comparison of PDC and PAC in the decomposition of toluene, using a TiO_2/γ-Al_2O_3/nickel foam catalyst in 2010 improved performance in the PAC model (Jiang et al., 2016; Li et al., 2015). While the toluene removal efficiency of PDC and PAC were nearly identical at 94% and 96%, respectively (compared to 74% for the plasma reaction alone without catalyst, all at maximum energy density). The air-fed plasma reaction produces O_3 as a result of the interaction between electrons and oxygen molecules; in the reaction without catalyst, O_3 concentrations increased with energy density. The introduction of the catalyst within the plasma reactor significantly reduced but could not eliminate the O_3, as some O_3 would be produced immediately prior to cycling out of the reactor. Placement of the catalyst post-reactor allowed all O_3 produced to interact with the catalyst and thus be reduced to undetectable levels (Jiang et al., 2016; Li et al., 2015).

7.3   OTHER CATALYSIS FACTORS

The location of the catalyst and the reactions between the catalyst and any generated species is one consideration for plasma catalysis. For example, plasma-reaction-produced O_3 was found to negatively interact with a $Pd/LaMnO_3$ catalyst for the oxidation of trichloroethylene (TCE) (Vandenbroucke et al., 2016). Here, Vandenbroucke et al. found mixed results using a PAC configuration, as the catalyst functioned to break down O_3 in addition to the targeted TCE, resulting in byproducts that were more toxic than TCE itself. This was accompanied by an expected decrease in O_3 itself as a byproduct of the catalyst reaction. The toxic byproducts of VOC decomposition via nonthermal plasma have been long noted, originally suggested as a method of removal for those compounds (Mizuno, 2007; Urashima and Chang, 2000). The Cu and Ce species as modified by the CAP discharge were determined to negatively interact and affect the long-term catalytic behavior, resulting in decreased performance. The combination of these studies suggests that the effect of the plasma-generated radicals on various catalysts will impact the resulting efficiencies of the reactions. In these cases, and many plasma catalysis reactions, the gaseous or vapor subjects are mixed with air (synthetic or natural) for injection into the plasma reactor. Thus, understanding the exact nature of radicals produced with an air-fed plasma reactor and the conditions that affect the production of those radicals is paramount for optimizing plasma-catalysis. For example, for VOC removal, in particular, increased O_2 presence in the feed gas led to increased catalytic efficiency for most but not all catalysts, with the relative surface area of the catalyst being one of the factors involved in the increase (Jiang et al., 2013a; Jwa et al., 2013; Kim et al., 2015a; Lee et al., 2015; Zhu et al., 2017a).

Toluene, as a representative VOC pollutant, widely exists in both indoor environments and industrial processes (Krishnamurthy et al., 2019; Ma et al., 2019). In addition to its decomposition through PDC or PAC, toluene also has a decomposition pathway using CAP and its reactive species as the catalysts directly. Figure 7.3a demonstrates possible pathways of toluene destruction by reactive species generated by CAP. Toluene destruction via cold plasma can generally be achieved through three pathways: electron-impacts, radicals attacks, and ion collisions (Lee and Chang, 2003; Zhu et al., 2017b). One of the most important pathways should be electron impact, which also should be responsible for the initial reaction for toluene destruction (Lee and Chang, 2003). Beyond energetic electrons' impact, the radical attack is another important mechanism resulting into toluene destruction. Gas-phase radicals, such as atomic O and •OH, are also deeply involved in toluene destruction, especially in subsequent oxidation after the opening of the aromatic ring of toluene. In addition, formic/acetic acid and benzaldehyde are the main intermediates in the photocatalytic oxidation of toluene where •OH is considered the primary strong oxidant (d'Hennezel et al., 1998; Irokawa et al., 2006; Mo et al., 2009a, 2009b; Zhao and Yang, 2003). No benzene is found in the photocatalytic process, while much benzene has been produced in cold plasma. In contrast, when using a traditional catalyst, toluene decomposition likewise had a preference for Mn oxide augmentation when compared to augmentation with Co oxides (Subrahmanyam et al., 2010). Similarly, Patil et al. found that increasing the loading of metal oxide support catalysts in a DBD beyond 5% reduced NO_x formation in the reactor, suggesting that both surface micro-discharges and thermal oxidation play roles in the catalysis (Patil et al., 2016). Further, surface-absorbed oxygen plays an important role in many catalysis reactions and has been determined to be important in plasma catalysis (Zhu et al., 2017b). In one case, it was determined that the electron spin resonance of silver and iron ions as governed by Si/Al ratio within a catalyst affected the catalytic performance of a reaction in oxygen plasma (Kim et al., 2015b).

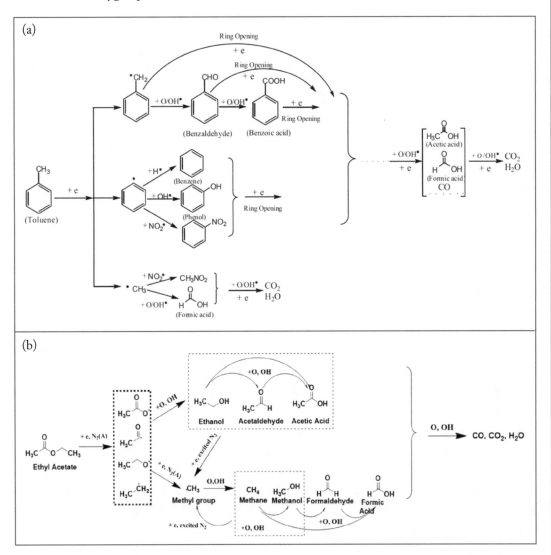

Figure 7.3: Possible major reaction pathways in the plasma catalysis. (a) Possible pathways of toluene destruction via reactive species generated by the cold plasma. Reproduced from (Huang et al., 2011). (b) Potential major reaction pathways in the plasma-directed removal of ethyl acetate. Reproduced from (Zhu et al., 2017b).

Ethyl acetate can be handled similarly to toluene. Figure 7.3b illustrates major reaction pathways in the plasma oxidation of ethyl acetate. Its initial decomposition pathways are via breaking of the C-C and C-O bonds, forming acetate ($CH_3COO\bullet$), acetyl ($CH_3CO\bullet$), ethyl ($CH_3CH2\bullet$), ethoxide ($CH_3CH_2O\bullet$), and methyl ($CH_3\bullet$) groups (Nimlos et al., 1996). In the plasma environment, most of the C_2 groups would be further decomposed to C_1 species including $CH_3\bullet$, $CH_2O\bullet$, CO, and CO_2 by ROS/RNS or electrons (Sun et al., 2012a). Both C_2 and C_1 groups would go through a series of oxidation reactions via collisions with ROS generated by the plasma. As a result, a proportion of intermediates and by-products are oxidized to the final products including CO, CO_2 and H_2O (Neyts, 2016). Another factor to consider is catalyst composition, while important for most catalysis, which is especially important in plasma-catalysis due to the nature of radicals produced by the discharge. As discussed previously, the interaction of radicals or molecules created by the discharge can have both beneficial and deleterious effects on catalytic efficiency depending on the specifics of the reaction (Kim et al., 2015a; Lee et al., 2015; Vandenbroucke et al., 2016). An additional study analyzing methanol decomposition over Mn-Ce catalysts found that concentrations of less than 50% Mn resulted in increased dispersal of the Mn ions on the surface of the catalyst and, as a result, increased oxygen mobility for the redox cycles (Zhu et al., 2016). Another study found the oxidation state of Mn directly affects ozone production in plasma catalysis (Pangilinan et al., 2016).

Duration of exposure to CAP, in the absence of secondary effects such as temperature changes, appears to increase efficiency linearly for some reactions with plateaus at certain time periods depending on the volume of subject to be catalyzed, while others show no such plateauing. For example, TCE decomposition over MnO_x metal fibers showed that residence time within the reactor did not significantly change the results of the reaction for a sufficiently small electrode gap (Magureanu et al., 2007). Exposure time must be balanced against overall energy efficiency, however, as increased exposure time means the reduced total flow of the subject through the reactor. In addition, some catalysts show improved efficiencies as exposure time increases, such as vanadium oxide in TCE decomposition (Oda, 2003). Thus, efficiency measures generally take exposure time into account and recommend a length of time that is high but not the highest tested, such as 50 mL/min flow recommended by Zeng et al. (2015) for dry methane reformation in a DBD reactor with Al_2O_3-supported metal catalysts. In contrast, NO and SO_2 decomposition was found to linearly increase with exposure duration until total decomposition was achieved.

Another variable is the presence of water vapor in the plasma reaction. Karuppiah et al. (2012a) found that water vapor presence positively affected nitrobenzene decomposition in the presence of a MnO_x catalyst via increased hydroxide ion production. Jiang et al. (2016) also found that the presence of a small percentage of water vapor (relative humidity peak of 25%) improved catalytic decomposition of benzene, though any additional vapor quickly reduced performance. In addition, one study found that increased humidity, while decreasing the primary methanol decomposition catalytic reaction, also decreased the prevalence of undesirable byproducts in what may be a beneficial trade-off for certain reactions (Norsic et al., 2018). Other factors in plasma catalysis are perhaps more obvious but vary with the nature of the reaction. For example, ammonia synthesis over Ru catalysts was found to depend on the frequency and applied voltage for the CAP discharge as well as the exposure time (Peng et al., 2016).

7.4   SUMMARY AND FUTURE DIRECTIONS

CAP-enabled catalysis is an important area of research where minor advancements may yield significant benefits (Neyts et al., 2015). Notably, the potential of CAP to directly recharge catalysts or otherwise improve catalytic reaction should be investigated, as even small improvements in efficiency can have major economic benefits for industrial-scale applications. Efficiency improvements for these types of plasma-assisted catalysis configurations may be best achieved through simple trial-and-error testing, as the basic reactions involved in the processes are well-enough understood without complex modeling (Barboun et al., 2019; Puliyalil et al., 2018; Sultana et al., 2015). Further, pretreatment of a sample or post-treatment of a catalyst to improve overall efficiency represents a relatively simple modification to most catalytic processes that might be considered. The complex interactions of reactant, catalyst, and plasma discharge make an overarching model of plasma catalysis difficult to develop (Kim et al., 2015a; Liu et al., 2002; Nozaki and Okazaki, 2013). Even when studying a single catalytic reaction, such as toluene removal, slight changes to the experimental setup seem to have large effects on resulting efficiencies or byproducts. The variety of experimental plasma configurations, including changes in ambient atmospheric conditions as well as voltage, current, discharge configuration, exposure time, etc., is likely a major contributor to these varying results. Isolation of the effects of specific parameters, such as water vapor content, will need to occur before the reasonable prediction of the outcome of a specific application can be made.

CHAPTER   8

Plasma Energy

The growing deployment of renewable energy sources like solar and wind presents a unique challenge for modern energy production: these power sources are either cyclical or irregular in nature, leading to wide swings in production capacity that may not be entirely predictable (Ahmad and Tahar, 2014; Fennell, 2019; Stougie et al., 2018). For example, solar power systems generally follow predictable diurnal and annual cycles but are subject to changes in cloud cover throughout the day while wind farms are subject to stochastic changes daily and seasonally (Howland et al., 2019; Jung et al., 2019; Siniscalchi-Minna et al., 2019; Xu et al., 2018c). Overproduction and networked grids across larger areas can help alleviate but not totally solve these issues, and challenges increase dramatically when the penetration of these renewable resources is beyond ~20% of the overall generation capacity. To manages these challenges, significant advancements and investments in energy storage technologies are needed (Larcher and Tarascon, 2015). While more familiar energy storage mechanisms such as chemical batteries are necessary, their manufacturing environmental cost can be prohibitively high. Other electric storage options such as pumped hydro, compressed air, electric-to-electric and solar thermal storage, and chemical storage options represent potentially more economical and environmentally-friendly options for energy storage (Rodríguez-Monroy et al., 2018; Sternberg and Bardow, 2015). If the chemical storage molecules are produced from environmental carbon or waste products, these chemicals have the added advantage of being net-zero carbon footprint in their reaction. The role of atmospheric plasma in energy storage focuses primarily around two areas: (1) the use of CAP in the creation or consumption of chemical storage mediums; and (2) nonthermal modification of various materials for use as physical components in electrical storage technologies (Bogaerts and Neyts, 2018; Dou et al., 2018). Additionally, CAP can be considered for reducing the overall energy requirements for industrial processes, production, and manufacturing.

8.1   CHEMICAL STORAGE

Non-fossil-fuel forms of chemical energy are constantly being sought as both reserve storage for power production as well as portable replacements for fossil fuels like gasoline. Multiple different gases are being considered with various challenges and opportunities for each. When these gases can be produced using renewable energy sources from atmospheric carbon or waste products high in carbon, the process of burning them for power production then becomes carbon-neutral (Gutknecht and Charles, 2019). Figure 8.1a reveals CAP as an emerging technology for energy storage.

Plasma technology is well known for gas conversion applications, such as CO_2 conversion into value-added chemicals or renewable fuels and N_2 fixation from the air to be used for the production of small building blocks, e.g., mineral fertilizer (Bogaerts and Neyts, 2018). CAP is generated by electric power and can easily be switched on/off, which makes it suitable for using intermittent renewable electricity (Perna et al., 2016; Yin et al., 2019). The production of synthesis gas (syngas, H_2:CO mixture) using CAP is one area being studied. While conversion of CH_4 and CO_2 into syngas is typically accomplished at higher temperatures, Ozkan et al. analyzed the production in a multi-electrode DBD reactor with moderate success (Ozkan et al., 2015). Of particular interest, the choice of feed gas had a dramatic effect on the selectivity of output: while CH_4 conversion is always higher than CO_2, the electron energy distributions for the argon and helium plasmas changed the ratios significantly. De Bie et al. (2015) extensively detailed the reaction mechanisms involved, noting that the final ratios of products could be controlled by careful manipulation of various parameters including residency time in the reactor.

Currently, energy efficiency for atmospheric pressure plasma-based CO_2 conversion is generally lower than the rates achieved by low-pressure microwave plasmas. Modeling by Berthelot and Bogaerts (2017) suggests that improved quenching after reaction of low-pressure plasma conversion products further improves conversion efficiency as pressures approach atmospheric. This premise has been further expanded on with the mixture of low-percentages of CO_2 or CO gas into an argon- or helium-fed RF plasma reactor, in which the noble gases appear to act to quench the excited CO_x molecules, leading to conversion efficiencies up to 85% at low CO_2 percentages (Stewig et al., 2020). Further expansion of CO_2 conversion at atmospheric pressures is thus likely within the next decade.

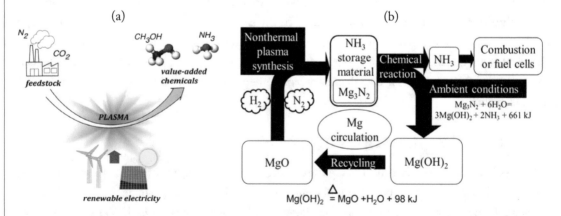

Figure 8-1: CAP for energy storage. (a) Plasma technology for gas conversions into value-added chemicals or renewable fuels for energy storage. Reproduced from (Bogaerts and Neyts, 2018). (b) Plasma-assisted Mg_3N_2 as a safe ammonia storage material. Reproduced from (Zen et al., 2019).

The storage and transport of chemical energy fuels is another area of study. For H_2 gas and hydrogen carriers like ammonia, storage in an inert medium is important for safety. One approach is to utilize metals such as magnesium as intermediate molecular "storage" for the relevant reactants. Figure 8.1b investigates improvement in the use of magnesium nitride as intermediate storage for ammonia by application of CAP (Zen et al., 2019). Mg_3N_2 can be generated by the nitridation of MgO with a cold plasma in an N_2 atmosphere (Zen et al., 2018). As the material that stores ammonia is synthesized directly without using ammonia in the proposed scheme, it will be more economical. Ammonia can later be synthesized in ambient conditions by exposure of the resulting magnesium nitride to sufficient water, with the magnesium oxide byproduct recycled back into storage use. For hydrogen, graphene sheets impregnated with magnesium via DBD proved to have significantly elevated kinetics with decreased working temperature when compared to pure magnesium storage (Lang et al., 2018). Importantly, the Mg layer was shown to be deposited uniformly on the few-layer graphene nanosheets.

Ammonia is considered another alternative fuel, but production difficulties need to be addressed to increase the feasibility of its use (Goto et al., 2017). In particular, the Haber-Bosch process primarily used for ammonia production requires high pressure (200–400 atm) and temperatures (400–600°C), with the result that current ammonia production is solely responsible for 2% of global energy use (Peng et al., 2018b). CAP, especially in conjunction with catalysis, offers multiple potential avenues for ammonia production without the energy-intensive conditions of Haber-Bosch (Mehta et al., 2018). The best CAP-assisted method for ammonia production was at approximately one-tenth of the production efficiency using Haber-Bosch; this finding indicates significant work needs to be done on the topic to approach industrial levels of efficiency (Hong et al., 2018). CAP may have a role in the decomposition of ammonia as well, as hydrogen gas is another prospect for chemical storage. Production of clean hydrogen gas from ammonia using non-thermal plasma was found to vary with applied power and resident time within the reactor, with 100% hydrogen production achieved at ambient temperatures and pressures requiring 50 W and 5 kV (Akiyama et al., 2018).

8.2   DIRECT ELECTRICAL STORAGE/PRODUCTION

Direct electrical storage technologies such as batteries, capacitors, and fuel cell technology also benefit from various applications of CAP. The specific treatments fall under surface modifications as discussed previously, mostly in the categories of functionalization and deposition. CAP processes have been applied to the manufacturing of electrodes and separation membranes. Different types of electrodes benefit from different treatments. Lithium metal oxide cathodes for lithium ion batteries have been treated, improving the retention of capacity for higher cycle times with several specific metals (Ajayi et al., 2017). Treatment of $Li_4Ti_5O_{12}$ electrodes for lithium ion batteries with an argon/N_2 feed gas increased conductivity and lithium ion diffusion (Lin et al., 2018b). Fuel cells differ from traditional batteries in that they produce electricity rather than store it, primary by reactions between hydrogen and oxygen (Das et al., 2017). Fuel cell electrodes utilize embedded catalysts, with CAP deposition of Pt or C subjects of study for efficiency improvement of the oxygen reduction reaction (Dou et al., 2018). Figure 8.2a displays a difference in the amount of waste with and without plasma jet treatment and cycling test/coulombic efficiency. Repeated plasma treatment resulted in the enhancement of electrodes and indicated a potential interaction between the electrode and the Li-N matrix electrolyte, while untreated electrodes had no such enhancement (de Guzman et al., 2014; Yang et al., 2014). Plasma jet treatment led to improved capacity retention through N-drop in the electrode, rearranging organic bonds, and SiO_2 oxidation (Guo et al., 2011).

CAP has also been used in conjunction with microbial fuel cell (MFC) technology. In MFCs, electrochemically-active bacteria are used for the current generation, often being fed biowaste products (Rahimnejad et al., 2015). A feasibility study by Chang et al. (2016) determined that an MFC containing an APPJ-treated carbon felt electrode exhibited superior performance than one without the modified electrode. PVC membranes for an alternate fuel cell technology, polymer electrolyte membrane (PEM) fuel cells, have shown improved performance through graft polymerizing acrylic acid prior to chemical grafting (Abu-Saied et al., 2019). Flow batteries are a hybrid between standard electrochemical batteries and fuel cells, where energy is both stored in cells and generated by cycling electrolyte fluid through the system (Minke and Turek, 2018). Application of an N_2 plasma jet to carbon felt electrodes for use in all-vanadium redox flow batteries improved overall efficiency of the battery from 62–76%, with the increase attributed to oxygen-containing functional groups and nitrogen doping on the surface (Chen et al., 2015b; Kazak et al., 2017). Multiple compositions of separator membranes have also been treated under atmospheric plasma for improved performance. Polyethylene separators for lithium ion batteries were treated with glow-like plasma discharge, resulting in improved cycling performance (Qin et al., 2018).

Small supercapacitors operate as power storage in many wearable electronics. Figure 8.2b demonstrates the reactive plasma coating process on a polyethylene (PE) separator and the resulting discharge rate capacity. Cells including PE separators coated with SiOxCyHz for an appropriate time exhibited improvements in charge transfer resistances, C rates and cycling performances (Qin et al., 2018). Polar groups such as Si(CH3)x, COOH, Si-OH, and C-OH were introduced by CAP into the PE separator, causing higher surface energy (Han et al., 2016a; Huang et al., 2013; Jin et al., 2017). The manufacturing of SnO_2/CNT supercapacitors using a nitrogen plasma jet has also been examined at various plasma exposure times, with shorter times associated with higher wettability and increased performance of the materials (Xu et al., 2016a). Fe_2O_3/CNT composites for supercapacitor use were likewise sintered using a nitrogen plasma jet by the authors with similar results (Xu et al., 2016b). In a similar study, vertical graphene tubes, which naturally exhibit poor capacitance due to low wettability, were treated in situ with an oxygen-fed plasma jet to increase hydrophilicity and, in turn, capacitance (Sahoo et al., 2018). While the capacitance gains were temporary and reverted with ambient exposure, the conducted tests provide a model of capacitance vs. wettability for graphene that can be used in future experiments. Further, the exact nature of the degradation of wettability was reported to depend on the specific functional groups created on the graphene surfaces, which may be a subject for future investigation toward the possible long-term stability of the capacitance. Graphene fiber-based supercapacitors showed remarkable improvement in general performance after the graphene fibers were treated with atmospheric plasma in ambient conditions (Meng et al., 2018). The resulting capacitors exhibited 33.1% increased areal capacitance, improved recycle stability, and other improvements when compared to capacitors made with the untreated fibers. Figure 8.2c illustrates a facile but robust strategy to easily enhance electrochemical properties of all-solid-state graphene fiber-based supercapacitors (GFSCs) by simple plasma treatment. As mentioned previously, plasma treatment can introduce oxygen-containing groups into materials such as graphene fibers, driving the transition from hydrophobicity to hydrophilicity. Such treatment can significantly impact the electrode/electrolyte interface, lowering resistance, and easing ion transport (Salanne et al., 2016). Separately, fiber-shaped supercapacitors were created from CAP-treated reduced graphene oxide fibers for use in AC line filtering (Zhao et al., 2019). The resulting capacitors featured higher conductivity due to the reduction in the concentration of oxygen-containing groups on the surface of the fibers. Such FSSCs are a valuable component for wearable electronics. The uses of cold plasmas in energy storage thus focus the synthesis of chemical fuels such as ammonia or and the creation or treatment of electrodes and separator membranes for existing battery technology and supercapacitors for small-scale electronic devices. There may also be the potential for manufacturing new lattices for the storage of chemical fuels.

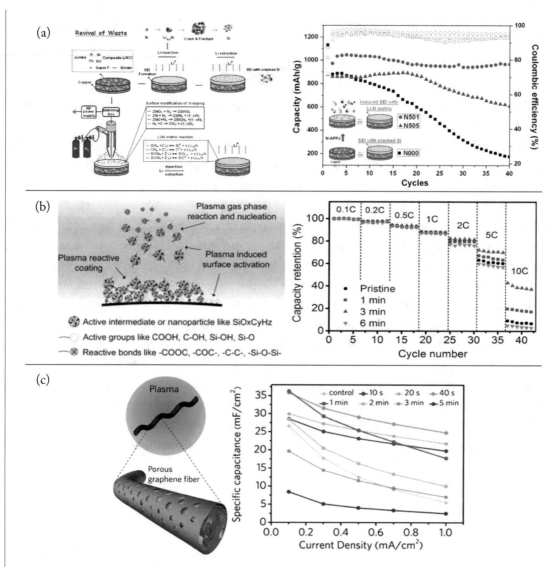

Figure 8.2: CAP for electrical storage. (a) Nitrogen APPJ-enhanced jumbo silicon/silicon carbide composite for lithium-ion batteries. Reproduced from (Chen et al., 2015a). (b) Binder-free nanoparticulate coating of a polyethylene separator via a reactive atmospheric pressure plasma for lithium-ion batteries with improved performance. Reproduced from (Qin et al., 2018). (c) Plasma treatment enhanced electrochemical performance of graphene fiber-based supercapacitors. Reproduced from (Meng et al., 2018).

8.3   SUMMARY AND FUTURE DIRECTIONS

In considering energy applications of CAP, the most obvious and immediate potential benefit could be in improving or replacing the Haber-Bosch process for ammonia production. Even a moderate reduction in the energy required for the process would have a profound impact on worldwide energy consumption. Such a reduction would most likely happen through reducing the energy required to break the nitrogen triple bond without allowing for the immediate decomposition of the resulting species; Haber-Bosch relies on high temperatures and pressures to achieve these results, but technology such as plasma-driven catalysis may provide alternatives at lower pressures and temperatures. Carreon proposes that some of the difficulty may be the current use of catalysts associated with thermal or neutral processes, whereas the reactions present in plasma-driven catalysis are different and should drive a search for new catalysts more appropriate for the specifics of CAP (Carreon, 2019). To date, attempts at replacing Haber-Bosch have yet to demonstrate the necessary efficiency to be an overall improvement, though understanding of the underlying plasma reactions is still in its infancy. Similarly, with the continued improvement of understanding in plasma for industrial applications, there are many energy-intensive processes that could reduce their footprint through judicious use of CAP.

With the broad and varied use of batteries in modern life, plasma-related advances in electrical storage is an area with high potential. Increases in energy density and reductions in recharge time are needed for many battery applications, especially transportation. Improvements to membranes and electrodes, especially with the elimination of rare earth metals or toxic components, are most likely to show economic and societal benefits. For example, aqueous electrolytes may help reduce the toxicity and cost of Li-ion batteries; CAP-treated cellulose-based Li-ion electrodes formed the basis of an experimental aqueous Li-ion battery with improved stability and sustainability, with the performance in aqueous electrolyte comparable to untreated electrodes in organic electrolyte (Profili et al., 2020). Other avenues, such as the development of large-scale, efficient redox flow batteries, might allow for a new model of personal transportation. For example, one could envision cars "filling up" with electrolyte (recycled on-site at the equivalent of gas stations) rather than gasoline to enable a quick method of renewably refueling vehicles for longer-distance travel. Gas conversion technologies also pose a significant potential, such as the conversion of ethanol to hydrogen gas via CAP, with byproduct carbon bound in "carbon dots" rather than becoming an additional environmental hazard (Zhou et al., 2020b).

CHAPTER 9

Conclusions, Challenges, and Future Work

9.1 OVERVIEW/CONCLUSION

Recent advances in CAP research, in particular and increasingly over the past two decades, show that CAP is a rapidly developing research field that promises not only many exciting scientific discoveries but also the development of new technologies and applications (Adamovich et al., 2017; Samukawa et al., 2012). CAP benefits multiple applications spanning from traditional disinfection in hygiene to more recent applications in agriculture, medicine, environment, materials, catalysis, energy, and so on (Bogaerts and Neyts, 2018; Bourke et al., 2018; Fridman et al., 2008; Kim et al., 2018; Kostov et al., 2014; Neyts et al., 2015). As in most fields of science (such as physics, chemistry, biology, and engineering), progress in applications has been ahead of the elementary understanding of the involved processes. With no doubt, CAP is itself a complex field of research with a wide range of research frontiers related to topics including plasma sources, plasma physics, plasma chemistry, diagnostics, plasma material interactions, material science, chemistry, and many other experimental/theoretical/computational research topics (Dey et al., 2016; Levchenko et al., 2018; Lu et al., 2016; Lu and Ostrikov, 2018; Vanraes and Bogaerts, 2018; Von Woedtke et al., 2013).

For agriculture, it is possible to harness the efficiency of plasma technology for different applications within different stages of food production (seeds, growth, harvest/packaging, transportation, market, home storage, meal preparation, and disposal/trash) by improving system design for enhancement of microbiological, physiological, and chemical quality characteristics of different types of foods (Gavahian and Cullen, 2020; Liao et al., 2020; Misra et al., 2019b; Niemira, 2012). Moreover, CAP inducing the modification of chemical structures within crops/foods has been elucidated and offers an avenue for adding value to byproducts waste streams, discovery of functional properties of crops/foods, and safety in terms of reducing immunoreactivity. CAP brings new and novel ways in agriculture to: (1) enhance germination, growth, and crop yields; (2) save water during cultivation; (3) protect crops; (4) sanitize/process food; and so on (Adhikari et al., 2020; Alves Filho et al., 2020; Misra et al., 2016b; Yodpitak et al., 2019). To boost real agricultural productivity in the near future, collaborations between plasma scientists, agricultural experts, plant biologists, and food technologists will be needed to demonstrate the mechanisms underlying plasma agriculture.

For medicine, applications of CAP such as disinfection, wound healing, blood coagulation, cancer therapies, dental disease treatment, and others are in the early but promising stages of their development (Laroussi et al., 2012; Lu et al., 2019; Von Woedtke et al., 2013). ROS and RNS are considered to be the key players in plasma-induced biological effects since ROS and RNS are components of regular physiological processes and, consequently, can interact with the processes of regular cell metabolism (Graves, 2012; Szili et al., 2018). Recent research has shown CAP to have potential to be non-invasively delivered, combined with other drugs, and used as a standalone drug (Chen et al., 2020a). There are also multiple opportunities for collaboration and convergence with biological and biomedical sciences including, but not limited to, clinical and preventative medicine, molecular biology, and biochemistry. CAP is establishing a new field of "plasma medicine" linking the interdisciplinary research works of plasma technology with biology and medicine, but the results obtained so far are still at a preliminary stage. CAP can alter biological systems through some combination of reactive species, electrons, ions, photons, and mechanical stresses, which need more work to clarify the mechanisms. More research is required to assess the toxicity and safety of CAP to humans and to clarify the mechanisms of interaction between plasma and living tissues. Although clinical application of CAP is growing rapidly, long-term side effects must be evaluated through clinical studies and the interaction mechanisms must be better understood. Thus, scientists need to continued clinical and scientific studies to understand CAP's mechanisms for clinical applications and reach its full potential for curing malignant diseases, especially cancers.

For the environment, CAP is a versatile processing technology with various applications in air, water, and soil (Aggelopoulos et al., 2020; Taranto et al., 2007; Yehia et al., 2020). Significant research efforts remain focused on improving the efficiency of these processes. Especially for volatile organic compounds, CAP can be an important tool for converting CO_2, CH_4, and syngas into value-added products, such as hydrocarbons, due to its unique ability to selectively initiate a variety of processes (Heijkers et al., 2020). CAP will play a significant role in solving challenging environmental problems. Scientists will make plasma science more accessible and widely understood by focusing on environmental problems from the air, water, and soil.

For materials, CAP has shown great potential in the areas of surface modification and material synthesis (Dou et al., 2018; Dufay et al., 2020; Macedo et al., 2020). Surface modifications include etching, cleaning, irradiation, functionalization of surface molecules, and deposition of thin films, layers, and nanoparticles. Synthesis includes bulk synthesis, as of gases, and synthesis of nanoparticles including carbon nanostructures and metal nanoparticles as used in catalysts and electrode manufacturing. Plasma material specialists and engineers will provide both the basic tools for scientific experimentation as well as develop new devices based on a better understanding of cold plasma physics. General strategies for plasma-assisted synthesis and surface modification will be proposed due to different types of plasma resulting into different effects on the materials (Merche et al., 2012; Van Deynse et al., 2016). CAP will be a more efficient strategy to synthesize/

modify materials when combining with other technologies (such as light, laser, thermo, etc.) and novel approaches will be envisioned to enrich CAP.

For catalysis, applications of CAP have been shown to improve catalysis in broad terms, with specific benefits resulting from specific classes of use (Bogaerts et al., 2020; Di et al., 2018). The efficiency of many solid catalysts is dependent upon factors such as total surface energy, specific molecular features of the surface, and the total surface area available for interaction, with CAP affecting all three beneficially from the standpoint of catalytic reactions (Kortshagen et al., 2016; Li et al., 2019a). CAP treatment of subjects, especially in vapor or gaseous form, has also been shown to be an effective method for accelerating desired reactions primarily by transferring energy from the discharge to the subject. In addition, radicals in the discharge (generated through energetic reactions with either the subject itself or the gas it is mixed with) may generate secondary and even tertiary reactions that can further improve synthesis or decomposition. With an increase in the number of viable commercial and industrial applications, it is also reasonable to expect that future research will focus on specific mechanisms affecting the efficiencies of these systems, primarily around minimizing energy investment for specific results. The free radicals and energy of the plasma discharge may regenerate catalysts for extended use and improved efficiency while the reaction is taking place (Chen et al., 2009b; Neyts et al., 2015). Thus, to eliminate the various sources of interaction in an experimental setup, multiple distinct configurations will be tested. Given the general role of catalysis in decreasing costs associated with many reactions and industrial processes, increased investment in PDC and PAC research is likely to occur, especially for reactions currently requiring high temperatures or nonstandard atmospheres. Efforts to develop catalysts with lower reaction energies or increased reaction efficiency, in general, will likely focus on specific surface properties of the catalysts and supports in question.

For energy, the utilization of CAP currently focuses on the synthesis of chemical fuels such as ammonia, the creation or treatment of electrodes and separator membranes for existing battery technology, and supercapacitors for small-scale electronic devices (Bogaerts and Neyts, 2018; Dou et al., 2018; Hong et al., 2018). There may also be the potential for manufacturing new lattices for the storage of chemical fuels. CAP can play an important role in the future energy infrastructure as it has great potential in combination with renewable energies for storage or use of peak energies and stabilization of the energy grid (Ajayi et al., 2017). Given the periodic nature of most renewable energy sources, large-scale energy storage systems that mimic the power density of fossil fuels will be needed to manage the off-production periods; redox flow batteries may provide such a system.

9.2   CAP FUNDAMENTAL RESEARCH (ACROSS APPLICATIONS AND CROSS-DISCIPLINARY)

The large number of examples discussed in this book suggest that CAP and its generation of reactive species represent an exciting rapidly expanding and evolving multidisciplinary field of research. The multidisciplinary characteristics of CAP introduce numerous and significant challenges that in turn open new opportunities. CAP research is "multi-phase" as it involves all four states of matter, for example, a propagating streamer head in the plasma state, reactive species in the gas phase, interacting solutions/media in the liquid state, and living tissue in the solid state (Lu et al., 2019, 2016). Reactive species generated in CAP and delivered to surfaces and even into the inner space of subjects are extremely complex and require detailed studies of physical, chemical, and biological phenomena and elementary processes. Relatively limited knowledge of CAP in certain areas will pose numerous challenges and new opportunities for future research and development (Adamovich et al., 2017). It brings synergistic studies where new knowledge will be obtained by employing experimental and theoretical approaches simultaneously. Due to being very non-uniform in space and varying with time over relatively short time scales, present-day CAP diagnostics generally characterize the averaged values of the species concentrations. As a result, most of the current numerical models are limited by the spatial and temporal resolution of diagnostic results. Thus, new and novel diagnostic methods, approaches, and instruments are urgently needed for in-situ monitoring of plasma processes and accurate capture of higher spatiotemporal resolution needs to obtain a more thorough understanding of this technology.

Each of the areas of CAP research also presents unique multdisciplinary research challenges. For agricultural applications of CAP, limited knowledge of the ongoing CAP-mediated effects within foods/crops has emerged, relating to seeds, plant tissues, insects, food safety, and so on. This reveals the need to understand the biochemical interactions in detail in each whole system to optimize the plasma processes for agriculture. Meanwhile, a full assessment is necessary for the benefits and risks of standalone CAP unit processes or their integration into a processing chain and what the ecological, economic, and consumer benefits and acceptability may be. To date, clinical research in plasma medicine has been mainly focused on applications such as dermatology, aesthetic surgery, and tumors. While the applications of CAP in the strategic area of human health still have large margins of improvements. Notably, *in vivo/in vitro* testing of CAP treatments is needed to better understand the effects of plasma-generated ROS/RNS and other species on biomolecules, cells, biological liquids, and tissues, and for the full development of plasma-based therapies. Apart from the usual safety concerns, a major obstacle exists due to the lack of fundamental understanding of biological, chemical, and physical mechanisms of interaction between CAP and living cells, tissues, organs, and the whole organism. During and after exposure to plasma-induced oxidative stresses, challenges for medicine and agriculture encompass three areas: (1) intracellular signal and response; (2) extracellular signal and stresses; and (3) membranes (cell walls) (Bradu et al., 2020; Graves, 2014c; Szili et al., 2018). For environmental applications of CAP, process selectivity, plasma delivery, conversion, and energy efficiency are still insufficient, and the challenge remains to justify the large-scale development. Developing CAP applications for air, water, and soil requires overcoming the following challenges: (1) improve large scale plasma processes with high energy efficiency, and (2) understand characteristic of CAP and accompanying mechanisms of chemical reactions (Bali et al., 2019; Kim, 2004; Liao et al., 2019). For material applications of CAP, the plasma process is extremely complicated and challenging to accurately reach the predesigned target of synthesizing and modifying the materials (Dou et al., 2018). The industrial-scaled applications of CAP in materials are also one of the most universally encountered issues in CAP research, which need to improve the utilization efficiency of plasma. Few theoretical research efforts on cold plasma materials ware described in the literature, which shows the need for a deeper understanding of the structure and the property changes after treatment by plasma. For catalysis applications of CAP, the complex mechanisms are far from understood (Bogaerts et al., 2020). Interactions of plasma catalysis are either the catalyst affecting the discharge or plasma changing the surface properties of catalysis by electrons, ions, reactive species, or photon interactions. For energy applications of CAP, the potential for significant cost savings through bypassing or replacing the Haber-Bosch process for ammonia production should drive further investment (Hong et al., 2018). Li-ion batteries as well are ubiquitous in modern society, and improvements in cycle time, lifespan, and charge density should be considered to increased profits and reduced pollution (Qin et al., 2018). Solving these challenges requires developing an understanding of the fundamental plasma characteristics in an integrated approach that combines new computational strategies and experimental (diagnostic) techniques.

To address some of the many aforementioned challenges, there are many ongoing projects and opportunities allowing the investigation of CAP. Scientists will develop new/novel diagnostics and numerical models to understand CAP. Plasma specialists will improve the utilization efficiency of CAP and reduce the waste of generated plasma. People will consider combining CAP technology and solar energy to lower its energy consumption and apply solar energy as power for plasma generation. More and deeper work will be needed to develop low cost, energy efficiency, and suitable facilities to simplify plasma technology. Plasma dose (delivery) will require more fully understanding the mechanisms of plasma therapeutics.

All in all, more CAP sources will be developed with low cost and high efficiency; diagnostics and numerical models will be proposed to understand mechanisms in plasma physics, plasma agriculture, plasma medicine, plasma environment, plasma material, plasma catalysis, plasma energy, and so on; large-scale and commercial plasma tools/equipment will be developed for industrial applications; and CAP clinical applications will become more possible. As a final statement, this work would not be possible without the efforts of thousands of researchers across dozens of fields. While we have done our best to identify a comprehensive sample of that body of work, we apologize if we have overlooked any particular efforts or contributions. Our goal is to stimulate an increase in collaborative research and development as well as encourage broad discussions on the subject of cold atmospheric plasma. Any comments, suggestions, or efforts for future information exchange are welcomed and appreciated.

References

Abd El-Aziz MF, Mahmoud EA, Elaragi GM (2014) Non thermal plasma for control of the Indian meal moth, Plodia interpunctella (Lepidoptera: Pyralidae). *Journal of Stored Products Research* 59: 215–221. DOI: 10.1016/j.jspr.2014.03.002. 31

Abou Saoud W, Assadi AA, Kane A, Jung A-V, Le Cann P, Gerard A, Bazantay F, Bouzaza A, Wolbert D (2020) Integrated process for the removal of indoor VOCs from food industry manufacturing: Elimination of Butane-2, 3-dione and Heptan-2-one by cold plasma-photocatalysis combination. *Journal of Photochemistry and Photobiology A: Chemistry* 386: 112071. DOI: 10.1016/j.jphotochem.2019.112071. 84

Aboubakr HA, Gangal U, Youssef MM, Goyal SM, Bruggeman PJ (2016) Inactivation of virus in solution by cold atmospheric pressure plasma: identification of chemical inactivation pathways. *Journal of Physics D: Applied Physics* 49: 204001. DOI: 10.1088/0022-3727/49/20/204001. 63

Aboubakr HA, Parra FS, Collins J, Bruggeman P, Goyal SM (2020) In situ inactivation of human norovirus GII. 4 by cold plasma: Ethidium monoazide (EMA)-coupled RT-qPCR underestimates virus reduction and fecal material suppresses inactivation. *Food Microbiology* 85: 103307. DOI: 10.1016/j.fm.2019.103307. 30

Abu-Saied M, Fahmy A, Morgan N, Qutop W, Abdelbary H, Friedrich JF (2019) Enhancement of poly (vinyl chloride) electrolyte membrane by its exposure to an atmospheric dielectric barrier discharge followed by grafting with polyacrylic acid. *Plasma Chemistry and Plasma Processing* 39: 1499–1517. DOI: 10.1007/s11090-019-10017-6. 98

Abuzairi T, Okada M, Bhattacharjee S, Nagatsu M (2016) Surface conductivity dependent dynamic behaviour of an ultrafine atmospheric pressure plasma jet for microscale surface processing. *Applied Surface Science* 390: 489–496. DOI: 10.1016/j.apsusc.2016.08.047. 71, 72

Acero JL, Stemmler K, Von Gunten U (2000) Degradation kinetics of atrazine and its degradation products with ozone and OH radicals: a predictive tool for drinking water treatment. *Environmental Science and Technology* 34: 591–597. DOI: 10.1021/es990724e. 67

Adachi T, Tanaka H, Nonomura S, Hara H, Kondo S-i, Hori M (2015) Plasma-activated medium induces A549 cell injury via a spiral apoptotic cascade involving the mitochondrial–nuclear network. *Free Radical Biology and Medicine* 79: 28–44. DOI: 10.1016/j.freeradbiomed.2014.11.014. 49, 50, 51

Adamovich I, Baalrud SD, Bogaerts A, Bruggeman P, Cappelli M, Colombo V, Czarnetzki U, Ebert U, Eden JG, Favia P (2017) The 2017 Plasma Roadmap: Low temperature *Plasma Science and Technology*. *Journal of Physics D: Applied Physics* 50: 323001. DOI: 10.1088/1361-6463/aa76f5. 6, 9, 13, 22, 71,103, 106

Adhikari B, Pangomm K, Veerana M, Mitra S, Park G (2020) Plant Disease Control by Non-Thermal Atmospheric-Pressure Plasma. *Frontiers in Plant Science* 11: 77. DOI: 10.3389/fpls.2020.00077. 103

Adnan Z, Mir S, Habib M (2017) Exhaust gases depletion using non-thermal plasma (NTP). *Atmospheric Pollution Research* 8: 338–343. DOI: 10.1016/j.apr.2016.10.005. 60

Aerts R, Somers W, Bogaerts A (2015) Carbon dioxide splitting in a dielectric barrier discharge plasma: a combined experimental and computational study. *ChemSusChem* 8: 702–716. DOI: 10.1002/cssc.201402818. 5

Aggelopoulos C, Gkelios A, Klapa M, Kaltsonoudis C, Svarnas P, Tsakiroglou C (2016) Parametric analysis of the operation of a non-thermal plasma reactor for the remediation of NAPL-polluted soils. *Chemical Engineering Journal* 301: 353–36. DOI: 10.1016/j.cej.2016.05.017. 67

Aggelopoulos C, Hatzisymeon M, Tatataki D, Rassias G (2020) Remediation of ciprofloxacin-contaminated soil by nanosecond pulsed dielectric barrier discharge plasma: Influencing factors and degradation mechanisms. *Chemical Engineering Journal* 124768. DOI: 10.1016/j.cej.2020.124768. 104

Aggelopoulos C, Svarnas P, Klapa M, Tsakiroglou C (2015a) Dielectric barrier discharge plasma used as a means for the remediation of soils contaminated by non-aqueous phase liquids. *Chemical Engineering Journal* 270: 428–436. DOI: 10.1016/j.cej.2015.02.056. 67

Aggelopoulos C, Tatataki D, Rassias G (2018) Degradation of atrazine in soil by dielectric barrier discharge plasma–Potential singlet oxygen mediation. *Chemical Engineering Journal* 347: 682–694. DOI: 10.1016/j.cej.2018.04.111. 67, 68

Aggelopoulos C, Tsakiroglou C, Ognier S, Cavadias S (2013) Ex situ soil remediation by cold atmospheric plasma discharge. *Procedia Environmental Sciences* 18: 649–656. DOI: 10.1016/j.proenv.2013.04.089. 67

Aggelopoulos C, Tsakiroglou C, Ognier S, Cavadias S (2015b) Non-aqueous phase liquid-contaminated soil remediation by ex situ dielectric barrier discharge plasma. *International Journal of Environmental Science and Technology* 12: 1011–1020. DOI: 10.1007/s13762-013-0489-4. 67

Ahmad S, Tahar RM (2014) Selection of renewable energy sources for sustainable development of electricity generation system using analytic hierarchy process: A case of Malaysia. *Renewable Energy* 63: 458–466. DOI: 10.1016/j.renene.2013.10.001. 95

Ahmed SF, Quadeer AA, McKay MR (2020) Preliminary identification of potential vaccine targets for the COVID-19 coronavirus (SARS-CoV-2) based on SARS-CoV immunological studies. *Viruses* 12: 254. DOI: 10.3390/v12030254. 56

Ajayi BP, Thapa AK, Cvelbar U, Jasinski JB, Sunkara MK (2017) Atmospheric plasma spray pyrolysis of lithiated nickel-manganese-cobalt oxides for cathodes in lithium ion batteries. *Chemical Engineering Science* 174: 302–310. DOI: 10.1016/j.ces.2017.09.022. 97, 105

Åkerström S, Gunalan V, Keng CT, Tan Y-J, Mirazimi A (2009) Dual effect of nitric oxide on SARS-CoV replication: viral RNA production and palmitoylation of the S protein are affected. *Virology* 395: 1–9. DOI: 10.1016/j.virol.2009.09.007. 55

Akishev Y, Grushin M, Karalnik V, Trushkin N, Kholodenko V, Chugunov V, Kobzev E, Zhirkova N, Irkhina I, Kireev G (2008) Atmospheric-pressure, nonthermal plasma sterilization of microorganisms in liquids and on surfaces. *Pure and Applied Chemistry* 80: 1953–1969. DOI: 10.1351/pac200880091953. 36

Akiyama M, Aihara K, Sawaguchi T, Matsukata M, Iwamoto M (2018) Ammonia decomposition to clean hydrogen using non-thermal atmospheric-pressure plasma. *International Journal of Hydrogen Energy* 43: 14493–14497. DOI: 10.1016/j.ijhydene.2018.06.022. 97

Al-Maliki H, Zsidai L, Samyn P, Szakál Z, Keresztes R, Kalácska G (2018) Effects of atmospheric plasma treatment on adhesion and tribology of aromatic thermoplastic polymers. *Polymer Engineering and Science* 58: E93–E103. DOI: 10.1002/pen.24689. 71

Alemán C, Fabregat G, Armelin E, Buendía JJ, Llorca J (2018) Plasma surface modification of polymers for sensor applications. *Journal of Materials Chemistry B* 6: 6515–6533. DOI: 10.1039/C8TB01553H. 71

Alotaibi F, Tung TT, Nine MJ, Kabiri S, Moussa M, Tran DN, Losic D (2018) Scanning atmospheric plasma for ultrafast reduction of graphene oxide and fabrication of highly conductive graphene films and patterns. *Carbon* 127: 113–121. DOI: 10.1016/j.carbon.2017.10.075. 76

Alshannaq A, Yu J-H (2017) Occurrence, toxicity, and analysis of major mycotoxins in food. *International Journal of Environmental Research and Public Health* 14: 632. DOI: 10.3390/ijerph14060632. 23

Alves Filho EG, de Brito ES, Rodrigues S (2020) Effects of cold plasma processing in food components. In: *Advances in Cold Plasma Applications for Food Safety and Preservation*, pp. 253–268. Elsevier: DOI: 10.1016/B978-0-12-814921-8.00008-6. 103

Ambardekar V, Bandyopadhyay P, Majumder S (2018) Atmospheric plasma sprayed SnO_2 coating for ethanol detection. *Journal of Alloys and Compounds* 752: 440–447. DOI: 10.1016/j.jallcom.2018.04.151. 71

Anbar M, Taube H (1954) Interaction of nitrous acid with hydrogen peroxide and with water. *Journal of the American Chemical Society* 76: 6243–6247. DOI: 10.1021/ja01653a007. 19

Andreozzi R, Caprio V, Insola A, Marotta R (1999) Advanced oxidation processes (AOP) for water purification and recovery. *Catalysis Today* 53: 51–59. DOI: 10.1016/S0920-5861(99)00102-9. 63

Apel K, Hirt H (2004) Reactive oxygen species: metabolism, oxidative stress, and signal transduction. *Annual Review Plant Biology* 55: 373–399. DOI: 10.1146/annurev.arplant.55.031903.141701. 29

Arik N, Inan A, Ibis F, Demirci EA, Karaman O, Ercan UK, Horzum N (2019) Modification of electrospun PVA/PAA scaffolds by cold atmospheric plasma: alignment, antibacterial activity, and biocompatibility. *Polymer Bulletin* 76: 797–812. DOI: 10.1007/s00289-018-2409-8. 71

Arndt S, Schmidt A, Karrer S, von Woedtke T (2018) Comparing two different plasma devices kINPen and Adtec SteriPlas regarding their molecular and cellular effects on wound healing. *Clinical Plasma Medicine* 9: 24–33. DOI: 10.1016/j.cpme.2018.01.002. 39, 56

Arndt S, Unger P, Wacker E, Shimizu T, Heinlin J, Li Y-F, Thomas HM, Morfill GE, Zimmermann JL, Bosserhoff A-K (2013) Cold atmospheric plasma (CAP) changes gene expression of key molecules of the wound healing machinery and improves wound healing *in vitro* and *in vivo*. *PloS One* 8: e79325. DOI: 10.1371/journal.pone.0079325. 39

Assadian O, Ousey KJ, Daeschlein G, Kramer A, Parker C, Tanner J, Leaper DJ (2019) Effects and safety of atmospheric low-temperature plasma on bacterial reduction in chronic wounds and wound size reduction: A systematic review and meta-analysis. *International Wound Journal* 16: 103–111. DOI: 10.1111/iwj.12999. 39

Astafan A, Batiot-Dupeyrat C, Pinard L (2019) Mechanism and Kinetic of Coke Oxidation by Nonthermal Plasma in Fixed-Bed Dielectric Barrier Reactor. *The Journal of Physical Chemistry C* 123: 9168–9175. DOI: 10.1021/acs.jpcc.9b00743. 71, 75

Atkinson R, Baulch D, Cox R, Hampson Jr R, Kerr J, Rossi M, Troe J (1997) Evaluated kinetic, photochemical and heterogeneous data for atmospheric chemistry: Supplement V.

IUPAC Subcommittee on Gas Kinetic Data Evaluation for Atmospheric Chemistry. *Journal of Physical and Chemical Reference Data* 26: 521–1011. DOI: 10.1063/1.556011. 15, 17

Attri P, Sarinont T, Kim M, Amano T, Koga K, Cho AE, Choi EH, Shiratani M (2015) Influence of ionic liquid and ionic salt on protein against the reactive species generated using dielectric barrier discharge plasma. *Scientific Reports* 5: 17781. DOI: 10.1038/srep17781. 5

Ayan H, Fridman G, Staack D, Gutsol AF, Vasilets VN, Fridman AA, Friedman G (2008) Heating effect of dielectric barrier discharges for direct medical treatment. *IEEE Transactions on Plasma Science* 37: 113–120. DOI: 10.1109/TPS.2008.2006899. 8

Aziz KHH, Miessner H, Mueller S, Kalass D, Moeller D, Khorshid I, Rashid MAM (2017) Degradation of pharmaceutical diclofenac and ibuprofen in aqueous solution, a direct comparison of ozonation, photocatalysis, and non-thermal plasma. *Chemical Engineering Journal* 313: 1033–1041. DOI: 10.1016/j.cej.2016.10.137. 64, 65

Babaeva NY, Naidis GV (2018) Modeling of plasmas for biomedicine. *Trends in Biotechnology* 36: 603–614. DOI: 10.1016/j.tibtech.2017.06.017. 14, 35

Babaie M, Davari P, Talebizadeh P, Zare F, Rahimzadeh H, Ristovski Z, Brown R (2015) Performance evaluation of non-thermal plasma on particulate matter, ozone and CO_2 correlation for diesel exhaust emission reduction. *Chemical Engineering Journal* 276: 240–248. DOI: 10.1016/j.cej.2015.04.086. 61

Bahrami N, Bayliss D, Chope G, Penson S, Perehinec T, Fisk ID (2016) Cold plasma: A new technology to modify wheat flour functionality. *Food Chemistry* 202: 247–253. DOI: 10.1016/j.foodchem.2016.01.113. 31

Bahri M, Haghighat F (2014) Plasma-B ased Indoor Air Cleaning Technologies: The State of the Art-R eview. *CLEAN–Soil, Air, Water* 42: 1667–1680. DOI: 10.1002/clen.201300296. 60

Bahri M, Haghighat F, Rohani S, Kazemian H (2016) Impact of design parameters on the performance of non-thermal plasma air purification system. *Chemical Engineering Journal* 302: 204–212. DOI: 10.1016/j.cej.2016.05.035. 60

Bai H, Zhang H (2017) Characteristics, sources, and cytotoxicity of atmospheric polycyclic aromatic hydrocarbons in urban roadside areas of Hangzhou, China. *Journal of Environmental Science and Health, Part A* 52: 303–312. DOI: 10.1080/10934529.2016.1258862. 66

Balakrishnan S, Mukherjee S, Das S, Bhat FA, Raja Singh P, Patra CR, Arunakaran J (2017) Gold nanoparticles–conjugated quercetin induces apoptosis via inhibition of EGFR/PI3K/Akt–mediated pathway in breast cancer cell lines (MCF-7 and MDA-MB-231). *Cell Biochemistry and Function* 35: 217–231. DOI: 10.1002/cbf.3266. 52

Balci B, Oturan N, Cherrier R, Oturan MA (2009) Degradation of atrazine in aqueous medium by electrocatalytically generated hydroxyl radicals. A kinetic and mechanistic study. *Water Research* 43: 1924–1934. DOI: 10.1016/j.watres.2009.01.021. 67

Bali N, Aggelopoulos C, Skouras E, Tsakiroglou C, Burganos V (2017) Hierarchical modeling of plasma and transport phenomena in a dielectric barrier discharge reactor. *Journal of Physics D: Applied Physics* 50: 505202. DOI: 10.1088/1361-6463/aa95fe. 67

Bali N, Aggelopoulos C, Skouras E, Tsakiroglou C, Burganos V (2019) Modeling of a DBD plasma reactor for porous soil remediation. *Chemical Engineering Journal* 373: 393–405. DOI: 10.1016/j.cej.2019.05.005. 107

Bang IH, Lee ES, Lee HS, Min SC (2020) Microbial decontamination system combining antimicrobial solution washing and atmospheric dielectric barrier discharge cold plasma treatment for preservation of mandarins. *Postharvest Biology and Technology* 162: 111102. DOI: 10.1016/j.postharvbio.2019.111102. 28

Barboun P, Mehta P, Herrera FA, Go DB, Schneider WF, Hicks JC (2019) Distinguishing plasma contributions to catalyst performance in plasma-assisted ammonia synthesis. *ACS Sustainable Chemistry and Engineering* 7: 8621–8630. DOI: 10.1021/acssuschemeng.9b00406. 93

Bárdos L, Baránková H (2010) Cold atmospheric plasma: Sources, processes, and applications. *Thin Solid Films* 518: 6705–6713. DOI: 10.1016/j.tsf.2010.07.044. 7, 35

Barrett K, McBride M (2005) Oxidative degradation of glyphosate and aminomethylphosphonate by manganese oxide. *Environmental Science and Technology* 39: 9223–9228. DOI: 10.1021/es051342d. 68

Bartis E, Graves D, Seog J, Oehrlein G (2013) Atmospheric pressure plasma treatment of lipopolysaccharide in a controlled environment. *Journal of Physics D: Applied Physics* 46: 312002. DOI: 10.1088/0022-3727/46/31/312002. 37

Bartis E, Knoll A, Luan P, Seog J, Oehrlein G (2016) On the interaction of cold atmospheric pressure plasma with surfaces of bio-molecules and model polymers. *Plasma Chemistry and Plasma Processing* 36: 121–149. DOI: 10.1007/s11090-015-9673-2. 71

Barton A, Wende K, Bundscherer L, Hasse S, Schmidt A, Bekeschus S, Weltmann K-D, Lindequist U, Masur K (2013) Nonthermal plasma increases expression of wound healing related genes in a keratinocyte cell line. *Plasma Medicine* 3. DOI: 10.1615/PlasmaMed.2014008540. 42

Bauer G (2016) The antitumor effect of singlet oxygen. *Anticancer Research* 36: 5649–5663. DOI: 10.21873/anticanres.11148. 52

Bauer G, Graves DB (2016) Mechanisms of selective antitumor action of cold atmospheric plasma-derived reactive oxygen and nitrogen species. *Plasma Processes and Polymers* 13: 1157–117. DOI: 10.1002/ppap.201600089. 35, 53

Beckman JS, Koppenol WH (1996) Nitric oxide, superoxide, and peroxynitrite: the good, the bad, and ugly. *American Journal of Physiology-cell Physiology* 271: C1424–C1437. DOI: 10.1152/ajpcell.1996.271.5.C1424. 45

Behrens M, Studt F, Kasatkin I, Kühl S, Hävecker M, Abild-Pedersen F, Zander S, Girgsdies F, Kurr P, Kniep B-L (2012) The active site of methanol synthesis over $Cu/ZnO/Al_2O_3$ industrial catalysts. *Science* 336: 893–897. DOI: 10.1126/science.1219831. 87

Bekeschus S, Brüggemeier J, Hackbarth C, von Woedtke T, Partecke L-I, van der Linde J (2017) Platelets are key in cold physical plasma-facilitated blood coagulation in mice. *Clinical Plasma Medicine* 7: 58–65. DOI: 10.1016/j.cpme.2017.10.001. 53, 55

Bekeschus S, Clemen R, Metelmann H-R (2018) Potentiating anti-tumor immunity with physical plasma. *Clinical Plasma Medicine* 12: 17–22. DOI: 10.1016/j.cpme.2018.10.001. 53

Bekeschus S, Clemen R, Nießner F, Sagwal SK, Freund E, Schmidt A (2020a) Medical gas plasma jet technology targets murine melanoma in an immunogenic fashion. *Advanced Science* 7: 1903438. DOI: 10.1002/advs.201903438. 56

Bekeschus S, Eisenmann S, Sagwal SK, Bodnar Y, Moritz J, Poschkamp B, Stoffels I, Emmert S, Madesh M, Weltmann K-D (2020b) xCT (SLC7A11) expression confers intrinsic resistance to physical plasma treatment in tumor cells. *Redox Biology* 30: 101423. DOI: 10.1016/j.redox.2019.101423. 35

Bekeschus S, Schmidt A, Weltmann K-D, von Woedtke T (2016) The plasma jet kINPen–A powerful tool for wound healing. *Clinical Plasma Medicine* 4: 19–28. DOI: /10.1016/j.cpme.2016.01.001. 41

Bernhardt T, Semmler ML, Schäfer M, Bekeschus S, Emmert S, Boeckmann L (2019) Plasma medicine: Applications of cold atmospheric pressure plasma in dermatology. *Oxidative Medicine and Cellular Longevity* 2019. DOI: 10.1155/2019/3873928. 39

Berthelot A, Bogaerts A (2017) Modeling of CO2 splitting in a microwave plasma: how to improve the conversion and energy efficiency. *The Journal of Physical Chemistry* C 121: 8236–8251. DOI: 10.1021/acs.jpcc.6b12840. 96

Bezza FA, Chirwa EMN (2017) The role of lipopeptide biosurfactant on microbial remediation of aged polycyclic aromatic hydrocarbons (PAHs)-contaminated soil. *Chemical Engineering Journal* 309: 563–576. DOI: 10.1016/j.cej.2016.10.055. 66

Bielski BH, Cabelli DE, Arudi RL, Ross AB (1985) Reactivity of HO2/O- 2 radicals in aqueous solution. *Journal of Physical and Chemical Reference Data* 14: 1041–1100. DOI: 10.1063/1.555739. 50

Birdsall N, Wheeler D (1993) Trade policy and industrial pollution in Latin America: where are the pollution havens? *The Journal of Environment and Development* 2: 137–149. DOI: 10.1177/107049659300200107. 60

Blanquet E, Nahil MA, Williams PT (2019) Enhanced hydrogen-rich gas production from waste biomass using pyrolysis with non-thermal plasma-catalysis. *Catalysis Today* 337: 216–224. DOI: 10.1016/j.cattod.2019.02.033. 89

Blin-Simiand N, Tardiveau P, Risacher A, Jorand F, Pasquiers S (2005) Removal of 2-Heptanone by Dielectric Barrier Discharges–The Effect of a Catalyst Support. *Plasma Processes and Polymers* 2: 256–262. DOI: 10.1002/ppap.200400088. 88

Blue E, Stanko J (1969) Pulsed langmuir probe measurements in a helium afterglow plasma. *Journal of Applied Physics* 40: 4061–4067. DOI: 10.1063/1.1657144. 3

Bocci V, Valacchi G, Corradeschi F, Aldinucci C, Silvestri S, Paccagnini E, Gerli R (1998) Studies on the biological effects of ozone: 7. Generation of reactive oxygen species (ROS) after exposure of human blood to ozone. *Journal of Biological Regulators and Homeostatic Agents* 12: 67–75. 53

Boehm D, Bourke P (2018) Safety implications of plasma-induced effects in living cells–a review of *in vitro* and *in vivo* findings. *Biological Chemistry* 400: 3–1. DOI: 10.1515/hsz-2018-0222. 42

Bogaerts A, Neyts EC (2018) Plasma technology: an emerging technology for energy storage. *ACS Energy Letters* 3: 1013–1027. DOI: 10.1021/acsenergylett.8b00184. 22, 95, 96, 103 105

Bogaerts A, Tu X, Whitehead JC, Centi G, Lefferts L, Guaitella O, Azzolina-Jury F, Kim H-H, Murphy AB, Schneider WF (2020) The 2020 plasma catalysis roadmap. *Journal of Physics D: Applied Physics* 53: 443001. DOI: 10.1088/1361-6463/ab9048. 105, 107

Bogle MA, Arndt KA, Dover JS (2007) Evaluation of plasma skin regeneration technology in low-energy full-facial rejuvenation. *Archives of Dermatology* 143: 168–174. DOI: 10.1001/archderm.143.2.168. 54

Bolouki N, Hsieh J-H, Li C, Yang Y-Z (2019) Emission spectroscopic characterization of a helium atmospheric pressure plasma jet with various mixtures of argon gas in the presence and the absence of de-ionized water as a target. *Plasma* 2: 283–293. DOI: 10.3390/plasma2030020. 12

Bond GC (1991) Supported metal catalysts: some unsolved problems. *Chemical Society Reviews* 20: 441-475. DOI: 10.1039/cs9912000441. 22, 80

Borena BM, Martens A, Broeckx SY, Meyer E, Chiers K, Duchateau L, Spaas JH (2015) Regenerative skin wound healing in mammals: state-of-the-art on growth factor and stem cell based treatments. *Cellular Physiology and Biochemistry* 36: 1–23. DOI: 10.1159/000374049. 39

Bormashenko E, Grynyov R, Bormashenko Y, Drori E (2012) Cold radiofrequency plasma treatment modifies wettability and germination speed of plant seeds. *Scientific Reports* 2: 1–8. DOI: 10.1038/srep00741. 24

Bormashenko E, Shapira Y, Grynyov R, Whyman G, Bormashenko Y, Drori E (2015) Interaction of cold radiofrequency plasma with seeds of beans (Phaseolus vulgaris). *Journal of Experimental Botany* 66: 4013–4021. DOI: /10.1093/jxb/erv206. 24

Bourke P, Ziuzina D, Boehm D, Cullen PJ, Keener K (2018) The potential of cold plasma for safe and sustainable food production. *Trends in Biotechnology* 36: 615–626. DOI: 10.1016/j.tibtech.2017.11.001. 6, 22, 23, 27, 32, 103

Bourke P, Ziuzina D, Han L, Cullen P, Gilmore B (2017) Microbiological interactions with cold plasma. *Journal of Applied Microbiology* 123: 308–324. DOI: 10.1111/jam.13429. 36

Bradu C, Kutasi K, Magureanu M, Puač N, Živković S (2020) Reactive nitrogen species in plasma-activated water: generation, chemistry and application in agriculture. *Journal of Physics D: Applied Physics* 53: 223001. DOI: 10.1088/1361-6463/ab795a. 107

Braithwaite NSJ (2000) Introduction to gas discharges. *Plasma Sources Science and Technology* 9: 517. DOI: 10.1088/0963-0252/9/4/307. 3

Brandenburg R (2017) Dielectric barrier discharges: progress on plasma sources and on the understanding of regimes and single filaments. *Plasma Sources Science and Technology* 26: 053001. DOI: 10.1088/1361-6595/aa6426. 7

Brandenburg R, Bogaerts A, Bongers W, Fridman A, Fridman G, Locke BR, Miller V, Reuter S, Schiorlin M, Verreycken T (2019) White paper on the future of plasma science in environment, for gas conversion and agriculture. *Plasma Processes and Polymers* 16: 1700238. DOI: 10.1002/ppap.201700238. 33

Brandenburg R, Lange H, von Woedtke T, Stieber M, Kindel E, Ehlbeck J, Weltmann K-D (2009) Antimicrobial effects of UV and VUV radiation of nonthermal plasma jets. *IEEE Transactions on Plasma Science* 37: 877–883. DOI: 10.1109/TPS.2009.2019657. 76

Braun S, Hanselmann C, Gassmann MG, auf dem Keller U, Born-Berclaz C, Chan K, Kan YW, Werner S (2002) Nrf2 transcription factor, a novel target of keratinocyte growth factor action which regulates gene expression and inflammation in the healing skin wound. *Molecular and Cellular Biology* 22: 5492–5505. DOI: 10.1128/MCB.22.15.5492-5505.2002. 42

Brehmer F, Haenssle H, Daeschlein G, Ahmed R, Pfeiffer S, Görlitz A, Simon D, Schön M, Wandke D, Emmert S (2015) Alleviation of chronic venous leg ulcers with a hand-held dielectric barrier discharge plasma generator (PlasmaDerm® VU-2010): results of a monocentric, two-armed, open, prospective, randomized and controlled trial (NCT 01415622). *Journal of the European Academy of Dermatology and Venereology* 29: 148–155. DOI: 10.1111/jdv.12490. 39, 56

Bruch-Gerharz D, Fehsel K, Suschek C, Michel G, Ruzicka T, Kolb-Bachofen V (1996) A proinflammatory activity of interleukin 8 in human skin: expression of the inducible nitric oxide synthase in psoriatic lesions and cultured keratinocytes. *The Journal of Experimental Medicine* 184: 2007–2012. DOI: 10.1084/jem.184.5.2007. 41

Bruggeman P, Brandenburg R (2013) Atmospheric pressure discharge filaments and microplasmas: physics, chemistry and diagnostics. *Journal of Physics D: Applied Physics* 46: 464001. DOI: 10.1088/0022-3727/46/46/464001. 9

Bruggeman P, Leys C (2009) Non-thermal plasmas in and in contact with liquids. *Journal of Physics D: Applied Physics* 42: 053001. DOI: 10.1088/0022-3727/42/5/053001. 28, 64

Bruggeman PJ, Iza F, Brandenburg R (2017) Foundations of atmospheric pressure non-equilibrium plasmas. *Plasma Sources Science and Technology* 26: 123002. DOI: 10.1088/1361-6595/aa97af. 5, 13

Bruggeman PJ, Kushner MJ, Locke BR, Gardeniers JG, Graham W, Graves DB, Hofman-Caris R, Maric D, Reid JP, Ceriani E (2016) Plasma–liquid interactions: a review and roadmap. *Plasma Sources Science and Technology* 25: 053002. DOI: 10.1088/0963-0252/25/5/053002. 28, 38

Brunekreef B, Holgate ST (2002) Air pollution and health. *The Lancet* 360: 1233–1242. DOI: 10.1016/S0140-6736(02)11274-8. 60

Bružauskaitė I, Bironaitė D, Bagdonas E, Bernotienė E (2016) Scaffolds and cells for tissue regeneration: different scaffold pore sizes—different cell effects. *Cytotechnology* 68: 355–369. DOI: 10.1007/s10616-015-9895-4. 73

Busco G, Omran AV, Ridou L, Pouvesle J-M, Robert E, Grillon C (2019) Cold atmospheric plasma-induced acidification of tissue surface: visualization and quantification using agarose gel models. *Journal of Physics D: Applied Physics* 52: 24LT01. DOI: 10.1088/1361-6463/ab1119. 53

Bußler S, Herppich WB, Neugart S, Schreiner M, Ehlbeck J, Rohn S, Schlüter O (2015) Impact of cold atmospheric pressure plasma on physiology and flavonol glycoside profile of peas (Pisum sativum 'Salamanca'). *Food Research International* 76: 132–141. DOI: 10.1016/j. foodres.2015.03.045. 76

Butscher D, Van Loon H, Waskow A, von Rohr PR, Schuppler M (2016) Plasma inactivation of microorganisms on sprout seeds in a dielectric barrier discharge. *International Journal of Food Microbiology* 238: —232. DOI: 10.1016/j.ijfoodmicro.2016.09.006. 24

Buzea C, Pacheco II, Robbie K (2007) Nanomaterials and nanoparticles: sources and toxicity. *Biointerphases* 2: MR17–MR71. DOI: 10.1116/1.2815690. 80

Capitelli M, Colonna G, D'Ammando G, Hassouni K, Laricchiuta A, Pietanza LD (2017) Coupling of plasma chemistry, vibrational kinetics, collisional-radiative models and electron energy distribution function under non-equilibrium conditions. *Plasma Processes and Polymers* 14: 1600109. DOI: 10.1002/ppap.201600109. 13

Capitelli M, Ferreira CM, Gordiets BF, Osipov AI (2013) *Plasma Kinetics in Atmospheric Gases.* Springer Science and Business Media. 15, 17

Capocelli M, Joyce E, Lancia A, Mason TJ, Musmarra D, Prisciandaro M (2012) Sonochemical degradation of estradiols: incidence of ultrasonic frequency. *Chemical Engineering Journal* 210: 9–17. DOI: 10.1016/j.cej.2012.08.084. 63

Cardoso R, Belmonte T, Henrion G, Sadeghi N (2006) Influence of trace oxygen on He (2 3S) density in a He–O$_2$ microwave discharge at atmospheric pressure: behaviour of the time afterglow. *Journal of Physics D: Applied Physics* 39: 4178. DOI: 10.1088/0022-3727/39/19/009. 14

Carreon ML (2019) Plasma catalytic ammonia synthesis: state of the art and future directions. *Journal of Physics D: Applied Physics* 52: 483001. DOI: 10.1088/1361-6463/ab3b2c. 101

Ceriani E, Marotta E, Shapoval V, Favaro G, Paradisi C (2018) Complete mineralization of organic pollutants in water by treatment with air non-thermal plasma. *Chemical Engineering Journal* 337: 567–575. DOI: 10.1016/j.cej.2017.12.107. 65, 66

Chacon L, Chen G, Knoll DA, Newman C, Park H, Taitano W, Willert JA, Womeldorff G (2017) Multiscale high-order/low-order (HOLO) algorithms and applications. *Journal of Computational Physics* 330: 21–45. DOI: 10.1016/j.jcp.2016.10.069. 14

Chang JW, Kang SU, Shin YS, Seo SJ, Kim YS, Yang SS, Lee J-S, Moon E, Lee K, Kim C-H (2015) Combination of NTP with cetuximab inhibited invasion/migration of cetuximab-resistant OSCC cells: Involvement of NF-κB signaling. *Scientific Reports* 5: 18208. DOI: 10.1038/srep18208. 46, 47

Chang R, Lu H, Tian Y, Li H, Wang J, Jin Z (2020) Structural modification and functional improvement of starch nanoparticles using vacuum cold plasma. *International Journal of Biological Macromolecules* 145: 197–206. DOI: 10.1016/j.ijbiomac.2019.12.167. 71

Chang S-H, Liou J-S, Liu J-L, Chiu Y-F, Xu C-H, Chen B-Y, Chen J-Z (2016) Feasibility study of surface-modified carbon cloth electrodes using atmospheric pressure plasma jets for microbial fuel cells. *Journal of Power Sources* 336: 99–106. DOI: 10.1016/j.jpowsour.2016.10.058. 98

Charoux CM, Free L, Hinds LM, Vijayaraghavan RK, Daniels S, O'Donnell CP, Tiwari BK (2020) Effect of non-thermal plasma technology on microbial inactivation and total phenolic content of a model liquid food system and black pepper grains. *LWT* 118: 108716. DOI: 10.1016/j.lwt.2019.108716. 23

Chauvin J, Gibot L, Griseti E, Golzio M, Rols M-P, Merbahi N, Vicendo P (2019) Elucidation of *in vitro* cellular steps induced by antitumor treatment with plasma-activated medium. *Scientific Reports* 9: 1–11. DOI: 10.1038/s41598-019-41408-6. 46

Chen B-H, Chuang S-I, Liu W-R, Duh J-G (2015a) A revival of waste: Atmospheric pressure nitrogen plasma jet enhanced jumbo silicon/silicon carbide composite in lithium ion batteries. *ACS Applied Materials and Interfaces* 7: 28166–28176. DOI: 10.1021/acsami.5b05858. 71, 100

Chen F, Liu J, Cui Y, Huang S, Song J, Sun J, Xu W, Liu X (2016a) Stability of plasma treated superhydrophobic surfaces under different ambient conditions. *Journal of Colloid and Interface Science* 470: 221–228. DOI: 10.1016/j.jcis.2016.02.058. 74

Chen F, Song J, Liu Z, Liu J, Zheng H, Huang S, Sun J, Xu W, Liu X (2016b) Atmospheric pressure plasma functionalized polymer mesh: An environmentally friendly and efficient tool for oil/water separation. *ACS Sustainable Chemistry and Engineering* 4: 6828–6837. DOI: 10.1021/acssuschemeng.6b01770. 74, 75, 81

Chen G, Chen Z, Wen D, Wang Z, Li H, Zeng Y, Dotti G, Wirz RE, Gu Z (2020a) Transdermal cold atmospheric plasma-mediated immune checkpoint blockade therapy. *Proceedings of the National Academy of Sciences* 117: 3687–3692. DOI: 10.1073/pnas.1917891117. 44, 51, 52, 53, 56, 104

Chen G, Georgieva V, Godfroid T, Snyders R, Delplancke-Ogletree M-P (2016c) Plasma assisted catalytic decomposition of CO_2. *Applied Catalysis B: Environmental* 190: 115-124. DOI: 10.1016/j.apcatb.2016.03.009. 84

Chen G, Zhou M, Chen S, Chen W (2009a) The different effects of oxygen and air DBD plasma byproducts on the degradation of methyl violet 5BN. *Journal of Hazardous Materials* 172: 786–791. DOI: 10.1016/j.jhazmat.2009.07.067. 64, 65

Chen H, Mu Y, Hardacre C, Fan X (2020b) Integration of membrane separation with nonthermal plasma catalysis: A proof-of-concept for CO_2 capture and utilization. *Industrial and Engineering Chemistry Research* 59: 8202–8211. DOI: 10.1021/acs.iecr.0c01067. 69

Chen H, Mu Y, Shao Y, Chansai S, Xiang H, Jiao Y, Hardacre C, Fan X (2020c) Nonthermal plasma (NTP) activated metal–organic frameworks (MOFs) catalyst for catalytic CO2 hydrogenation. *AIChE Journal* 66: e16853. DOI: 10.1002/aic.16853. 69

Chen HL, Lee HM, Chen SH, Chang MB, Yu SJ, Li SN (2009b) Removal of volatile organic compounds by single-stage and two-stage plasma catalysis systems: a review of the performance enhancement mechanisms, current status, and suitable applications. *Environmental Science and Technology* 43: 2216–2227. DOI: 10.1021/es802679b. 105

Chen J-Z, Liao W-Y, Hsieh W-Y, Hsu C-C, Chen Y-S (2015b) All-vanadium redox flow batteries with graphite felt electrodes treated by atmospheric pressure plasma jets. *Journal of Power Sources* 274: 894–898. DOI: 10.1016/j.jpowsour.2014.10.097. 71, 98

Chen K, Xu D, Li J, Geng X, Zhong K, Yao J (2020d) Application of terahertz time-domain spectroscopy in atmospheric pressure plasma jet diagnosis. *Results in Physics* 16: 102928. DOI: 10.1016/j.rinp.2020.102928. 9

Chen Q, Ozkan A, Chattopadhyay B, Baert K, Poleunis C, Tromont A, Snyders R, Delcorte A, Terryn H, Delplancke-Ogletree M-P (2019) N-doped TiO_2 photocatalyst coatings synthesized by a cold atmospheric plasma. *Langmuir* 35: 7161–7168. DOI: 10.1021/acs.langmuir.9b00784. 86

Chen W-Y, Matthews A, Jones FR, Chen K-S (2018a) Immobilization of carboxylic acid groups on polymeric substrates by plasma-enhanced chemical vapor or atmospheric pressure plasma deposition of acetic acid. *Thin Solid Films* 666: 547161–60. DOI: 10.1016/j.tsf.2018.07.051. 71

Chen X, Porto CL, Chen Z, Merenda A, Allioux F-M, d'Agostino R, Magniez K, Dai XJ, Palumbo F, Dumée LF (2018b) Single step synthesis of Janus nano-composite membranes by atmospheric aerosol plasma polymerization for solvents separation. *Science of the Total Environment* 645: 227161–33. DOI: 10.1016/j.scitotenv.2018.06.343. 74

Chen Y, Wu F, Lin Y, Deng N, Bazhin N, Glebov E (2007) Photodegradation of glyphosate in the ferrioxalate system. *Journal of Hazardous Materials* 148: 3607161–365. DOI: 10.1016/j.jhazmat.2007.02.044. 68

Chen Y-Q, Cheng J-H, Sun D-W (2020e) Chemical, physical and physiological quality attributes of fruit and vegetables induced by cold plasma treatment: Mechanisms and application advances. *Critical Reviews in Food Science and Nutrition* 60: 26767161–2690. DOI: 10.1080/10408398.2019.1654429. 33

Chen Z, Cheng X, Lin L, Keidar M (2016d) Cold atmospheric plasma discharged in water and its potential use in cancer therapy. *Journal of Physics D: Applied Physics* 50: 015208. DOI: 10.1088/1361-6463/50/1/015208. 35, 36, 55

Chen Z, Garcia Jr G, Arumugaswami V, Wirz RE (2020f) Cold atmospheric plasma for SARS-CoV-2 inactivation. *Physics of Fluids* 32: 111702. DOI: 10.1063/5.0031332. 38

Chen Z, Lin L, Cheng X, Gjika E, Keidar M (2016e) Effects of cold atmospheric plasma generated in deionized water in cell cancer therapy. *Plasma Processes and Polymers* 13: 7161–1156. DOI: 10.1002/ppap.201600086. 44, 46

Chen Z, Lin L, Cheng X, Gjika E, Keidar M (2016f) Treatment of gastric cancer cells with nonthermal atmospheric plasma generated in water. *Biointerphases* 11: 031010. DOI: 10.1116/1.4962130. 54

Chen Z, Lin L, Gjika E, Cheng X, Canady J, Keidar M (2017a) Selective treatment of pancreatic cancer cells by plasma-activated saline solutions. *IEEE Transactions on Radiation and Plasma Medical Sciences* 2: 1167161–120. DOI: 10.1109/TRPMS.2017.2761192. 44, 45

Chen Z, Lin L, Zheng Q, Sherman JH, Canady J, Trink B, Keidar M (2018c) Micro-sized cold atmospheric plasma source for brain and breast cancer treatment. *Plasma Medicine* 8: 2037161–215. DOI: 10.1615/PlasmaMed.2018026588. 48

Chen Z, Obenchain R, Wirz RE (2021) Tiny cold atmospheric plasma jet for biomedical applications. *Processes* 9: 249. DOI: 10.3390/pr9020249. 5

Chen Z, Simonyan H, Cheng X, Gjika E, Lin L, Canady J, Sherman JH, Young C, Keidar M (2017b) A novel micro cold atmospheric plasma device for glioblastoma both *in vitro* and *in vivo*. *Cancers* 9: 61. DOI: 10.3390/cancers9060061. 44, 47, 48

Chen Z, Wirz R (2020) Cold atmospheric plasma for COVID-19. *Preprints* 2020, 2020040126 DOI: 10.20944/preprints202004.0126.v1. 37, 38

Chen Z, Zhang S, Levchenko I, Beilis II, Keidar M (2017c) *In vitro* demonstration of cancer inhibiting properties from stratified self-organized plasma-liquid interface. *Scientific Reports* 7: 17161–11. DOI: 10.1038/s41598-017-12454-9. 43

Cheng X, Sherman J, Murphy W, Ratovitski E, Canady J, Keidar M (2014) The effect of tuning cold plasma composition on glioblastoma cell viability. *PloS One* 9: e98652. DOI: 10.1371/journal.pone.0098652. 9

Chesnokova N, Gundorova R, Krvasha O, Bakov V, Davydova N (2003) Experimental investigation of the influence of gas flow containing nitrogen oxide in treatment of corneal wounds. *Vestnik RAMN* 5: 407161–44. 54

Choi S, Puligundla P, Mok C (2017) Effect of corona discharge plasma on microbial decontamination of dried squid shreds including physico-chemical and sensory evaluation. *Lebensmittel-Wissenschaft & Technologie* 75: 3237161–328. DOI: 10.1016/j.lwt.2016.08.063. 31

Chung CH, Parker JS, Ely K, Carter J, Yi Y, Murphy BA, Ang KK, El-Naggar AK, Zanation AM, Cmelak AJ (2006) Gene expression profiles identify epithelial-to-mesenchymal transition and activation of nuclear factor-κB signaling as characteristics of a high-risk head and neck squamous cell carcinoma. *Cancer Research* 66: 82107161–8218. DOI: 10.1158/0008-5472.CAN-06-1213. 47

Church D, Elsayed S, Reid O, Winston B, Lindsay R (2006) Burn wound infections. *Clinical Microbiology Reviews* 19: 4037161–434. DOI: 10.1128/CMR.19.2.403-434.2006. 39

Cialfi S, Oliviero C, Ceccarelli S, Marchese C, Barbieri L, Biolcati G, Uccelletti D, Palleschi C, Barboni L, De Bernardo C (2010) Complex multipathways alterations and oxidative stress are associated with Hailey–Hailey disease. *British Journal of Dermatology* 162: 5187161–526. DOI: 10.1111/j.1365-2133.2009.09500.x. 54

Clark SJ, Segall MD, Pickard CJ, Hasnip PJ, Probert MI, Refson K, Payne MC (2005) First principles methods using CASTEP. *Zeitschrift für Kristallographie-Crystalline Materials* 220: 5677161–570. DOI: 10.1524/zkri.220.5.567.65075. 87

Clarke D, Tyuftin AA, Cruz-Romero MC, Bolton D, Fanning S, Pankaj SK, Bueno-Ferrer C, Cullen PJ, Kerry JP (2017) Surface attachment of active antimicrobial coatings onto conventional plastic-based laminates and performance assessment of these materials on the storage life of vacuum packaged beef sub-primals. *Food Microbiology* 62: 1967161–201. DOI: 10.1016/j.fm.2016.10.022. 31

Coburn J, Chen M (1980) Optical emission spectroscopy of reactive plasmas: A method for correlating emission intensities to reactive particle density. *Journal of Applied Physics* 51: 3134–3136. DOI: 10.1063/1.328060. 12

Colvin VL (2003) The potential environmental impact of engineered nanomaterials. *Nature Biotechnology* 21: 1166–1170. DOI: 10.1038/nbt875. 80

Cordes C, Bornath T, Redmer R (2016) Monte Carlo simulation of partially ionized hydrogen plasmas. *Contributions to Plasma Physics* 56: 475–481. DOI: 10.1002/ctpp.201600005. 13

Coutinho NM, Silveira MR, Fernandes LM, Moraes J, Pimentel TC, Freitas MQ, Silva MC, Raices RS, Ranadheera CS, Borges FO (2019) Processing chocolate milk drink by low–pressure cold plasma technology. *Food Chemistry* 278: 276–283. DOI: 10.1016/j.foodchem.2018.11.061. 6

Coutinho NM, Silveira MR, Rocha RS, Moraes J, Ferreira MVS, Pimentel TC, Freitas MQ, Silva MC, Raices RS, Ranadheera CS (2018) Cold plasma processing of milk and dairy products. *Trends in Food Science and Technology* 74: 56–68. DOI: 10.1016/j.tifs.2018.02.008. 30

Critzer FJ, Kelly–Wintenberg K, South SL, Golden DA (2007) Atmospheric plasma inactivation of foodborne pathogens on fresh produce surfaces. *Journal of Food Protection* 70: 2290–2296. DOI: 10.4315/0362-028X-70.10.2290. 37

Cui S, Hao R, Fu D (2019) Integrated method of non–thermal plasma combined with catalytical oxidation for simultaneous removal of SO_2 and NO. *Fuel* 246: 365–374. DOI: 10.1016/j.fuel.2019.03.012. 62

Cullen PJ, Lalor J, Scally L, Boehm D, Milosavljević V, Bourke P, Keener K (2018) Translation of plasma technology from the lab to the food industry. *Plasma Processes and Polymers* 15: 1700085. DOI: 10.1002/ppap.201700085. 23, 31

Cvelbar U, Walsh JL, Černák M, de Vries HW, Reuter S, Belmonte T, Corbella C, Miron C, Hojnik N, Jurov A (2019) White paper on the future of *Plasma Science and Technology* in plastics and textiles. *Plasma Processes and Polymers* 16: 1700228. DOI: 10.1002/ppap.201700228. 71

Czaplicka M, Kaczmarczyk B (2006) Infrared study of chlorophenols and products of their photodegradation. *Talanta* 70: 940–949. DOI: 10.1016/j.talanta.2006.05.060. 67

d'Hennezel O, Pichat P, Ollis DF (1998) Benzene and toluene gas–phase photocatalytic degradation over H_2O and HCL pretreated TiO_2: by–products and mechanisms. *Journal of Photochemistry and Photobiology A: Chemistry* 118: 197–204. DOI: 10.1016/S1010-6030(98)00366-9. 90

Daeschlein G (2013) Antimicrobial and antiseptic strategies in wound management. *International Wound Journal* 10: 9–14. DOI: 10.1111/iwj.12175. 39

Dai X, Bazaka K, Richard DJ, Thompson ERW, Ostrikov KK (2018) The emerging role of gas plasma in oncotherapy. *Trends in Biotechnology* 36: 1183–1198. DOI: 10.1016/j.tibtech.2018.06.010. 43

Daiber A, Ullrich V (2002) Stickstoffmonoxid, Superoxid und Peroxynitrit: Radikalchemie im Organismus. *Chemie in unserer Zeit* 36: 366–375. DOI: 10.1002/1521-3781(200212)36:6<366::AID-CIUZ366>3.0.CO;2-B. 19

Daria S, Sindu S, Mathias H, Luka H, Janik M, Jannes D, Zaho K, Bodo F, Holger K, Rainer A (2018) Surface modification of highly porous 3D networks via atmospheric plasma treatment. *Contributions to Plasma Physics* 58: 384–393. DOI: 10.1002/ctpp.201700120. 72

Darny T, Pouvesle J–M, Fontane J, Joly L, Dozias S, Robert E (2017a) Plasma action on helium flow in cold atmospheric pressure plasma jet experiments. *Plasma Sources Science and Technology* 26: 105001. DOI: 10.1088/1361-6595/aa8877. 10

Darny T, Pouvesle JM, Puech V, Douat C, Dozias S, Robert E (2017b) Analysis of conductive target influence in plasma jet experiments through helium metastable and electric field measurements. *Plasma Sources Science and Technology* 26: 045008. DOI: 10.1088/1361-6595/aa5b15. 9

Das V, Padmanaban S, Venkitusamy K, Selvamuthukumaran R, Blaabjerg F, Siano P (2017) Recent advances and challenges of fuel cell based power system architectures and control: A review. *Renewable and Sustainable Energy Reviews* 73: 10–18. DOI: 10.1016/j.rser.2017.01.148. 98

De Bie C, van Dijk J, Bogaerts A (2015) The dominant pathways for the conversion of methane into oxygenates and syngas in an atmospheric pressure dielectric barrier discharge. *The Journal of Physical Chemistry C* 119: 22331–22350. DOI: 10.1021/acs.jpcc.5b06515. 96

de Groot GJ, Hundt A, Murphy AB, Bange MP, Mai–Prochnow A (2018) Cold plasma treatment for cotton seed germination improvement. *Scientific Reports* 8: 1–10. DOI: 10.1038/s41598-018-32692-9. 24

de Guzman RC, Yang J, Cheng MM–C, Salley SO, Ng KS (2014) High capacity silicon nitride-based composite anodes for lithium ion batteries. *Journal of Materials Chemistry A* 2: 14577–14584. DOI: 10.1039/C4TA02596B. 98

De Smet L, Vancoillie G, Minshall P, Lava K, Steyaert I, Schoolaert E, Van De Walle E, Dubruel P, De Clerck K, Hoogenboom R (2018) Plasma dye coating as straightforward and widely applicable procedure for dye immobilization on polymeric materials. *Nature Communications* 9: 1–11. DOI: 10.1038/s41467-018-03583-4. 71

Deng X–T, Shi J, Chen H, Kong MG (2007) Protein destruction by atmospheric pressure glow discharges. *Applied Physics Letters* 90: 013903. DOI: 10.1063/1.2410219. 9

Denifl S, Feil S, Winkler M, Probst M, Echt O, Matt–Leubner S, Grill V, Scheier P, Märk T (2008) Electron impact ionization/dissociation of hydrocarbon molecules relevant to the edge plasma. *Atomic and Plasma-Material Interaction Data for Fusion* (APID Series) 14: 1–11. 3

Dey A, Chroneos A, Braithwaite NSJ, Gandhiraman RP, Krishnamurthy S (2016) Plasma engineering of graphene. *Applied Physics Reviews* 3: 021301. DOI: 10.1063/1.4947188. 80, 103

Di L, Zhang J, Zhang X (2018) A review on the recent progress, challenges, and perspectives of atmospheric-pressure cold plasma for preparation of supported metal catalysts. *Plasma Processes and Polymers* 15: 1700234. DOI: 10.1002/ppap.201700234. 80

Dilecce G, Martini L, Tosi P, Scotoni M, De Benedictis S (2015) Laser induced fluorescence in atmospheric pressure discharges. *Plasma Sources Science and Technology* 24: 034007. DOI: 10.1088/0963-0252/24/3/034007. 10

Dimitrakellis P, Gogolides E (2018a) Atmospheric plasma etching of polymers: A palette of applications in cleaning/ashing, pattern formation, nanotexturing and superhydrophobic surface fabrication. *Microelectronic Engineering* 194: 109–115. DOI: 10.1016/j.mee.2018.03.017. 75

Dimitrakellis P, Gogolides E (2018b) Hydrophobic and superhydrophobic surfaces fabricated using atmospheric pressure cold plasma technology: A review. *Advances in Colloid and Interface Science* 254: 1–21. DOI: 10.1016/j.cis.2018.03.009. 71, 81

Dobrynin D, Fridman G, Friedman G, Fridman A (2009) Physical and biological mechanisms of direct plasma interaction with living tissue. *New Journal of Physics* 11: 115020. DOI: 10.1088/1367-2630/11/11/115020. 49

Dobrynin D, Wu A, Kalghatgi S, Park S, Chernets N, Wasko K, Dumani E, Ownbey R, Joshi SG, Sensenig R (2011) Live pig skin tissue and wound toxicity of cold plasma treatment. *Plasma Medicine* 1. DOI: 10.1615/PlasmaMed.v1.i1.80. 42

Dockery DW, Pope CA, Xu X, Spengler JD, Ware JH, Fay ME, Ferris Jr BG, Speizer FE (1993) An association between air pollution and mortality in six US cities. *New England Journal of Medicine* 329: 1753–1759. DOI: 10.1056/NEJM199312093292401. 60

Dong J, Wang Z, Zhao X, Biesuz M, Saunders T, Zhang Z, Hu C, Grasso S (2020) Contactless flash sintering based on cold plasma. *Scripta Materialia* 175: 20–23. DOI: 10.1016/j.scriptamat.2019.08.039. 71

Dorai R, Kushner MJ (2003) A model for plasma modification of polypropylene using atmospheric pressure discharges. *Journal of Physics D: Applied Physics* 36: 666. DOI: 10.1088/0022-3727/36/6/309. 18

Dou S, Tao L, Wang R, El Hankari S, Chen R, Wang S (2018) Plasma-assisted synthesis and surface modification of electrode materials for renewable energy. *Advanced Materials* 30: 1705850. DOI: 10.1002/adma.201705850. 6, 77, 81, 95, 98, 104, 105, 107

Dougherty TJ, Gomer CJ, Henderson BW, Jori G, Kessel D, Korbelik M, Moan J, Peng Q (1998) Photodynamic therapy. *JNCI: Journal of the National Cancer Institute* 90: 889–905. DOI: 10.1093/jnci/90.12.889. 49

Dubinov A, Lazarenko E, Selemir V (2000) Effect of glow discharge air plasma on grain crops seed. *IEEE Transactions on Plasma Science* 28: 180–183. DOI: 10.1109/27.842898. 24

Dubuc A, Monsarrat P, Virard F, Merbahi N, Sarrette J–P, Laurencin–Dalicieux S, Cousty S (2018) Use of cold–atmospheric plasma in oncology: a concise systematic review. *Therapeutic Advances in Medical Oncology* 10: 1–12. DOI: 10.1177/1758835918786475. 43

Dufay M, Jimenez M, Degoutin S (2020) Effect of cold plasma treatment on electrospun nanofibers properties: a review. *ACS Applied Bio Materials*. DOI: 10.1021/acsabm.0c00154. 71, 104

Duske K, Jablonowski L, Koban I, Matthes R, Holtfreter B, Sckell A, Nebe JB, von Woedtke T, Weltmann KD, Kocher T (2015) Cold atmospheric plasma in combination with mechanical treatment improves osteoblast growth on biofilm covered titanium discs. *Biomaterials* 52: 327–334. DOI: 10.1016/j.biomaterials.2015.02.035. 54

Ehlbeck J, Schnabel U, Polak M, Winter J, Von Woedtke T, Brandenburg R, Von dem Hagen T, Weltmann K (2010) Low temperature atmospheric pressure plasma sources for microbial decontamination. *Journal of Physics D: Applied Physics* 44: 013002. DOI: 10.1088/0022-3727/44/1/013002. 37

Ekezie F–GC, Sun D–W, Cheng J–H (2017) A review on recent advances in cold plasma technology for the food industry: Current applications and future trends. *Trends in Food Science and Technology* 69: 46–58. DOI: 10.1016/j.tifs.2017.08.007. 31, 33

Ericson B, Hanrahan D, Kong V (2008) *The World's Worst Pollution Problems: The Top Ten of the Toxic Twenty*. Blacksmith Institute and Green Cross. 60

Evan GI, Vousden KH (2001) Proliferation, cell cycle and apoptosis in cancer. *Nature* 411: 342–348. DOI: 10.1038/35077213. 35

Eykens L, De Sitter K, Paulussen S, Dubreuil M, Dotremont C, Pinoy L, Van der Bruggen B (2018) Atmospheric plasma coatings for membrane distillation. *Journal of Membrane Science* 554: 175–183. DOI: 10.1016/j.memsci.2018.02.067. 76

Fabregat G, Osorio J, Castedo A, Armelin E, Buendía JJ, Llorca J, Alemán C (2017) Plasma functionalized surface of commodity polymers for dopamine detection. *Applied Surface Science* 399: 638–647. DOI: 10.1016/j.apsusc.2016.12.137. 71, 73

Fang FC (2004) Antimicrobial reactive oxygen and nitrogen species: concepts and controversies. *Nature Reviews Microbiology* 2: 820–832. DOI: 10.1038/nrmicro1004. 41

Fathollah S, Mirpour S, Mansouri P, Dehpour AR, Ghoranneviss M, Rahimi N, Naraghi ZS, Chalangari R, Chalangari KM (2016) Investigation on the effects of the atmospheric pressure plasma on wound healing in diabetic rats. *Scientific Reports* 6: 19144. DOI: 10.1038/srep19144. 42

Felix T, Cassini F, Benetoli L, Dotto M, Debacher N (2017) Morphological study of polymer surfaces exposed to non–thermal plasma based on contact angle and the use of scaling laws. *Applied Surface Science* 403: 57–61. DOI: 10.1016/j.apsusc.2017.01.036. 71

Feng F, Ye L, Liu J, Yan K (2013) Non–thermal plasma generation by using back corona discharge on catalyst. *Journal of Electrostatics* 71: 179–184. DOI: 10.1016/j.elstat.2012.11.034. 86

Feng X, Liu H, He C, Shen Z, Wang T (2018) Synergistic effects and mechanism of a non–thermal plasma catalysis system in volatile organic compound removal: a review. *Catalysis Science and Technology* 8: 936–954. DOI: 10.1039/C7CY01934C. 61, 83

Fennell PS (2019) Comparative energy analysis of renewable electricity and carbon capture and storage. *Joule* 3: 1406–1408. DOI: 10.1016/j.joule.2019.06.003. 95

Filimonova E, Amirov RK (2001) Simulation of ethylene conversion initiated by a streamer corona in an air flow. Plasma Physics Reports 27: 708–714. DOI: 10.1134/1.1390542. 23

Filipić A, Gutierrez–Aguirre I, Primc G, Mozetič M, Dobnik D (2020) Cold plasma, a new hope in the field of virus inactivation. *Trends in Biotechnology*. DOI: 10.1016/j.tibtech.2020.04.003. 6

Fluhr JW, Sassning S, Lademann O, Darvin ME, Schanzer S, Kramer A, Richter H, Sterry W, Lademann J (2012) *In vivo* skin treatment with tissue-tolerable plasma influences skin physiology and antioxidant profile in human stratum corneum. *Experimental Dermatology* 21: 130–134. DOI: 10.1111/j.1600-0625.2011.01411.x. 42

Foest R, Bindemann T, Brandenburg R, Kindel E, Lange H, Stieber M, Weltmann KD (2007) On the vacuum ultraviolet radiation of a miniaturized non-thermal atmospheric pressure plasma jet. *Plasma Processes and Polymers* 4: S460–S464. DOI: 10.1002/ppap.200731207. 76

Foest R, Kindel E, Ohl A, Stieber M, Weltmann K (2005) Non–thermal atmospheric pressure discharges for surface modification. *Plasma Physics and Controlled Fusion* 47: B525–B536. DOI: 10.1088/0741-3335/47/12B/S38. 76

Foresti R, Clark JE, Green CJ, Motterlini R (1997) Thiol compounds interact with nitric oxide in regulating heme oxygenase–1 induction in endothelial cells Involvement of superoxide and peroxynitrite anions. *Journal of Biological Chemistry* 272: 18411–18417. DOI: 10.1074/jbc.272.29.18411. 55

Förstermann U, Sessa WC (2012) Nitric oxide synthases: regulation and function. *European Heart Journal* 33: 829–837. DOI: 10.1093/eurheartj/ehr304. 50

Foster KW, Moy RL, Fincher EF (2008) Advances in plasma skin regeneration. *Journal of Cosmetic Dermatology* 7: 169–179. DOI: 10.1111/j.1473-2165.2008.00385.x. 54, 55

France J, King M, Lee–Taylor J (2007) Hydroxyl (OH) radical production rates in snowpacks from photolysis of hydrogen peroxide (H_2O_2) and nitrate (NO_3^-). *Atmospheric Environment* 41: 5502–5509. DOI: 10.1016/j.atmosenv.2007.03.056. 20

Frank S, Hübner G, Breier G, Longaker MT, Greenhalgh DG, Werner S (1995) Regulation of vascular endothelial growth factor expression in cultured keratinocytes. Implications for normal and impaired wound healing. *Journal of Biological Chemistry* 270: 12607–12613. DOI: 10.1074/jbc.270.21.12607. 42

Freund E, Liedtke KR, van der Linde J, Metelmann H–R, Heidecke C–D, Partecke L–I, Bekeschus S (2019) Physical plasma–treated saline promotes an immunogenic phenotype in CT26 colon cancer cells *in vitro* and *in vivo*. *Scientific Reports* 9: 1–18. DOI: 10.1038/s41598-018-37169-3. 46

Fridman A (2008) *Plasma Chemistry*. Cambridge University Press. DOI: 10.1017/CBO9780511546075. 3

Fridman A, Chirokov A, Gutsol A (2005) Non–thermal atmospheric pressure discharges. *Journal of Physics D: Applied Physics* 38: R1. DOI: 10.1088/0022-3727/38/2/R01. 7, 49

Fridman G, Friedman G, Gutsol A, Shekhter AB, Vasilets VN, Fridman A (2008) Applied plasma medicine. *Plasma Processes and Polymers* 5: 503–533. DOI: 10.1002/ppap.200700154. 35, 54, 103

Fridman G, Shereshevsky A, Jost MM, Brooks AD, Fridman A, Gutsol A, Vasilets V, Friedman G (2007) Floating electrode dielectric barrier discharge plasma in air promoting apoptotic behavior in melanoma skin cancer cell lines. *Plasma Chemistry and Plasma Processing* 27: 163–176. DOI: 10.1007/s11090-007-9048-4. 9

Friedman PC, Miller V, Fridman G, Lin A, Fridman A (2017) Successful treatment of actinic keratoses using nonthermal atmospheric pressure plasma: a case series. *Journal of the American Academy of Dermatology* 76: 349–350. DOI: 10.1016/j.jaad.2016.09.004. 53

Fritz JL, Owen MJ (1995) Hydrophobic recovery of plasma–treated polydimethylsiloxane. *The Journal of Adhesion* 54: 33–45. DOI: 10.1080/00218469508014379. 74

Fröhling A, Durek J, Schnabel U, Ehlbeck J, Bolling J, Schlüter O (2012) Indirect plasma treatment of fresh pork: Decontamination efficiency and effects on quality attributes. *Innovative*

Food Science and Emerging Technologies 16: 381–390. DOI: 10.1016/j.ifset.2012.09.001. 31

Frykberg RG, Banks J (2015) Challenges in the treatment of chronic wounds. *Advances in Wound Care 4*: 560–582. DOI: 10.1089/wound.2015.0635. 39

Fumagalli F, Kylián O, Amato L, Hanuš J, Rossi F (2012) Low–pressure water vapour plasma treatment of surfaces for biomolecules decontamination. *Journal of Physics D: Applied Physics* 45: 135203. DOI: 10.1088/0022-3727/45/13/135203. 73

Gao J, Ma C, Xing S, Sun L (2017) Oxidation behaviours of particulate matter emitted by a diesel engine equipped with a NTP device. *Applied Thermal Engineering* 119: 593–602. DOI: 10.1016/j.applthermaleng.2017.03.101. 61

García–Alcantara E, López–Callejas R, Morales–Ramírez PR, Peña–Eguiluz R, Fajardo–Muñoz R, Mercado–Cabrera A, Barocio SR, Valencia–Alvarado R, Rodríguez–Méndez BG, Muñoz–Castro AE (2013) Accelerated mice skin acute wound healing *in vivo* by combined treatment of argon and helium plasma needle. *Archives of Medical Research* 44: 169–177. DOI: 10.1016/j.arcmed.2013.02.001. 41

Gauthier L, Morel A, Anceriz N, Rossi B, Blanchard–Alvarez A, Grondin G, Trichard S, Cesari C, Sapet M, Bosco F (2019) Multifunctional natural killer cell engagers targeting NKp46 trigger protective tumor immunity. *Cell* 177: 1701–1713. e1716. DOI: 10.1016/j.cell.2019.04.041. 53

Gavahian M, Cullen P (2020) Cold plasma as an emerging technique for mycotoxin–free food: efficacy, mechanisms, and trends. *Food Reviews International* 36: 193–214. DOI: 10.1080/87559129.2019.1630638. 103

Gebicki S, Gebicki JM (1999) Crosslinking of DNA and proteins induced by protein hydroperoxides. *Biochemical Journal* 338: 629–636. DOI: 10.1042/bj3380629. 49

Georghiou GE, Papadakis A, Morrow R, Metaxas A (2005) Numerical modelling of atmospheric pressure gas discharges leading to plasma production. *Journal of Physics D: Applied Physics* 38: R303. DOI: 10.1088/0022-3727/38/20/R01. 13

Gerland P, Raftery AE, Ševčíková H, Li N, Gu D, Spoorenberg T, Alkema L, Fosdick BK, Chunn J, Lalic N (2014) World population stabilization unlikely this century. *Science* 346: 234–237. DOI: 10.1126/science.1257469. 23

Gerling T, Brandenburg R, Wilke C, Weltmann K–D (2017) Power measurement for an atmospheric pressure plasma jet at different frequencies: distribution in the core plasma and the effluent. *The European Physical Journal Applied Physics* 78: 10801. DOI: 10.1051/epjap/2017160489. 12

Gerullis S, Pohle L, Pfuch A, Beier O, Kretzschmar B, Raugust M, Rädlein E, Grünler B, Schimanski A (2018) Structural, electrical and optical properties of SnO_x films deposited by use of atmospheric pressure plasma jet. *Thin Solid Films* 649: 97–105. DOI: 10.1016/j.tsf.2018.01.037. 71

Ghasemi M, Olszewski P, Bradley J, Walsh J (2013) Interaction of multiple plasma plumes in an atmospheric pressure plasma jet array. *Journal of Physics D: Applied Physics* 46: 052001. DOI: 10.1088/0022-3727/46/5/052001. 10

Giardina A, Schiorlin M, Marotta E, Paradisi C (2020) Atmospheric pressure non-thermal plasma for air purification: Ions and ionic reactions induced by dc+ corona discharges in air contaminated with acetone and methanol. *Plasma Chemistry and Plasma Processing* 40: 1091–1107. DOI: 10.1007/s11090-020-10087-x. 69

Gilmore BF, Flynn PB, O'Brien S, Hickok N, Freeman T, Bourke P (2018) Cold plasmas for biofilm control: opportunities and challenges. *Trends in Biotechnology* 36: 627–638. DOI: 10.1016/j.tibtech.2018.03.007. 53

Goldstein S, Czapski G (1995) The reaction of $NO\cdot$ with $O_2\cdot-$ and $HO_2\cdot-$: A pulse radiolysis study. *Free Radical Biology and Medicine* 19: 505–510. DOI: 10.1016/0891-5849(95)00034-U. 19

Golubovskii G, Zenger V, Nasedkin A, Prokofieva E, Borojtsov A (2004) Use of exogenic NO–therapy in regenerative surgery of bronchial tubes, *IVth Army Medical Conference on Intensive Therapy and Profilactic Treatments of Surgical Infections.* 54

Golubovskii YB, Maiorov V, Behnke J, Behnke J (2002) Modelling of the homogeneous barrier discharge in helium at atmospheric pressure. *Journal of Physics D: Applied Physics* 36: 39. DOI:10.1088/0022-3727/36/1/306. 14

Gómez–Ramírez A, Cotrino J, Lambert R, González–Elipe A (2015) Efficient synthesis of ammonia from N_2 and H_2 alone in a ferroelectric packed–bed DBD reactor. *Plasma Sources Science and Technology* 24: 065011. DOI: 10.1088/0963-0252/24/6/065011. 87

Gómez–Ramírez A, López–Santos C, Cantos M, García JL, Molina R, Cotrino J, Espinós J, González–Elipe AR (2017) Surface chemistry and germination improvement of Quinoa seeds subjected to plasma activation. *Scientific Reports* 7: 1–12. DOI: 10.1038/s41598-017-06164-5. 27

Gong X, Huang D, Liu Y, Zeng G, Wang R, Wan J, Zhang C, Cheng M, Qin X, Xue W (2017) Stabilized nanoscale zerovalent iron mediated cadmium accumulation and oxidative damage of Boehmeria nivea (L.) Gaudich cultivated in cadmium contaminated sediments. *Environmental Science and Technology* 51: 11308–11316. DOI: 10.1021/acs.est.7b03164. 66

Good RJ (1992) Contact angle, wetting, and adhesion: a critical review. *Journal of Adhesion Science and Technology* 6: 1269–1302. DOI: 10.1163/156856192X00629. 73

Goto Y, Hayakawa Y, Kambara S (2017) Reaction mechanism of ammonia decomposition by atmospheric plasma. *Ninth JSME-KSME Thermal and Fluids Engineering Conference*, Okinawa, Japan, October 28-30. 97

Graves DB (1989) Plasma processing in microelectronics manufacturing. *AIChE Journal* 35: 1–29. DOI: 10.1002/aic.690350102. 3

Graves DB (1994) Plasma processing. *IEEE Transactions on Plasma Science* 22: 31–42. DOI: 10.1109/27.281547. 3

Graves DB (2012) The emerging role of reactive oxygen and nitrogen species in redox biology and some implications for plasma applications to medicine and biology. *Journal of Physics D: Applied Physics* 45: 263001. DOI: 10.1088/0022-3727/45/26/263001. 35, 49, 53, 57, 104

Graves DB (2014a) Low temperature plasma biomedicine: A tutorial review. *Physics of Plasmas* 21: 080901. DOI: 10.1063/1.4892534. 35

Graves DB (2014b) Oxy–nitroso shielding burst model of cold atmospheric plasma therapeutics. *Clinical Plasma Medicine* 2: 38–49. DOI: 10.1016/j.cpme.2014.11.001. 35, 53

Graves DB (2014c) Reactive species from cold atmospheric plasma: implications for cancer therapy. *Plasma Processes and Polymers* 11: 1120–1127. DOI: 10.1002/ppap.201400068. 35, 107

Green JA, Kirwan JM, Tierney JF, Symonds P, Fresco L, Collingwood M, Williams CJ (2001) Survival and recurrence after concomitant chemotherapy and radiotherapy for cancer of the uterine cervix: a systematic review and meta–analysis. *The Lancet* 358: 781–786. DOI: 10.1016/S0140-6736(01)05965-7. 57

Griem HR (1963) Validity of local thermal equilibrium in plasma spectroscopy. *Physical Review* 131: 1170. DOI: 10.1103/PhysRev.131.1170. 4

Gristina R, D'Aloia E, Senesi GS, Milella A, Nardulli M, Sardella E, Favia P, d'Agostino R (2009) Increasing cell adhesion on plasma deposited fluorocarbon coatings by changing the surface topography. *Journal of Biomedical Materials Research Part B: Applied Biomaterials: An Official Journal of The Society for Biomaterials, The Japanese Society for Biomaterials, and The Australian Society for Biomaterials and the Korean Society for Biomaterials* 88: 139–149. DOI: 10.1002/jbm.b.31160. 42

Guan W–j, Ni Z–y, Hu Y, Liang W–h, Ou C–q, He J–x, Liu L, Shan H, Lei C–l, Hui DS (2020) Clinical characteristics of coronavirus disease 2019 in China. *New England Journal of Medicine* 382: 1708–1720. DOI: 10.1056/NEJMoa2002032. 37

Guild III GN, Runner RP, Castilleja GM, Smith MJ, Vu CL (2017) Efficacy of hybrid plasma scalpel in reducing blood loss and transfusions in direct anterior total hip arthroplasty. *The Journal of Arthroplasty* 32: 458–462. DOI: 10.1016/j.arth.2016.07.038. 7

Guo FH, De Raeve HR, Rice TW, Stuehr DJ, Thunnissen F, Erzurum SC (1995) Continuous nitric oxide synthesis by inducible nitric oxide synthase in normal human airway epithelium *in vivo*. *Proceedings of the National Academy of Sciences* 92: 7809–7813. DOI: 10.1073/pnas.92.17.7809. 55

Guo H, Chen J, Wang L, Wang AC, Li Y, An C, He J–H, Hu C, Hsiao VK, Wang ZL (2020) A highly efficient triboelectric negative air ion generator. *Nature Sustainability* 1–7. 22

Guo J, Huang K, Wang J (2015) Bactericidal effect of various non–thermal plasma agents and the influence of experimental conditions in microbial inactivation: *A Review. Food Control* 50: 482–490. DOI: 10.1016/j.foodcont.2014.09.037. 31

Guo J, Sun A, Chen X, Wang C, Manivannan A (2011) Cyclability study of silicon–carbon composite anodes for lithium–ion batteries using electrochemical impedance spectroscopy. *Electrochimica Acta* 56: 3981–3987. DOI: 10.1016/j.electacta.2011.02.014. 98

Guo Sa, DiPietro LA (2010) Factors affecting wound healing. *Journal of Dental Research* 89: 219–229. DOI: 10.1177/0022034509359125. 39

Gupta RK, Dubey M, Kharel P, Gu Z, Fan QH (2015) Biochar activated by oxygen plasma for supercapacitors. *Journal of Power Sources* 274: 1300–1305. DOI: 10.1016/j.jpowsour.2014.10.169. 71, 80

Gurtler JB, Keller SE (2019) Microbiological safety of dried spices. *Annual Review of Food Science and Technology* 10: 409–427. DOI: 10.1146/annurev-food-030216-030000. 31

Gurunathan S, Han JW, Park JH, Kim E, Choi Y–J, Kwon D–N, Kim J–H (2015) Reduced graphene oxide–silver nanoparticle nanocomposite: a potential anticancer nanotherapy. *International Journal of Nanomedicine* 10: 6257. DOI: 10.2147/IJN.S92449. 76

Gutknecht V, Charles L (2019) CO_2 Capture from Air: A Breakthrough Sustainable Carbon Source for Synthetic Fuels. In: *Zukünftige Kraftstoffe*, pp. 181–190. Springer. DOI: 10.1007/978-3-662-58006-6_10. 95

Haertel B, Von Woedtke T, Weltmann K–D, Lindequist U (2014) Non–thermal atmospheric–pressure plasma possible application in wound healing. *Biomolecules and Therapeutics* 22: 477. DOI: 10.4062/biomolther.2014.105. 39

Hammer T (2014) Atmospheric pressure plasma application for pollution control in industrial processes. *Contributions to Plasma Physics* 54: 187–201. DOI: 10.1002/ctpp.201310063. 59

Han M, Kim D–W, Kim Y–C (2016a) Charged polymer–coated separators by atmospheric plasma–induced grafting for lithium–ion batteries. *ACS Applied Materials and Interfaces* 8: 26073–26081. DOI: 10.1021/acsami.6b08781. 98

Han SH, Suh HJ, Hong KB, Kim SY, Min SC (2016b) Oral toxicity of cold plasma-treated edible films for food coating. *Journal of Food Science* 81: T3052–T3057. DOI: 10.1111/1750-3841.13551. 31

Han Y, Cheng J–H, Sun D–W (2019) Activities and conformation changes of food enzymes induced by cold plasma: *A Review. Critical Reviews in Food Science and Nutrition* 59: 794–811. DOI: 10.1080/10408398.2018.1555131. 31, 33

Hanna K, Chiron S, Oturan MA (2005) Coupling enhanced water solubilization with cyclodextrin to indirect electrochemical treatment for pentachlorophenol contaminated soil remediation. *Water Research* 39: 2763–2773. DOI: 10.1016/j.watres.2005.04.057. 66

Haralambiev L, Wien L, Gelbrich N, Lange J, Bakir S, Kramer A, Burchardt M, Ekkernkamp A, Gümbel D, Stope MB (2020) Cold atmospheric plasma inhibits the growth of osteosarcoma cells by inducing apoptosis, independent of the device used. *Oncology Letters* 19: 283–290. DOI: 10.3892/ol.2019.11115. 35

Harley JC, Suchowerska N, McKenzie DR (2020) Cancer treatment with gas plasma and with gas plasma–activated liquid: positives, potentials and problems of clinical translation. *Biophysical Reviews* 1–18. DOI: 10.1007/s12551-020-00743-z. 44

Hasse S, Hahn O, Kindler S, von Woedtke T, Metelmann H–R, Masur K (2014) Atmospheric pressure plasma jet application on human oral mucosa modulates tissue regeneration. *Plasma Medicine* 4. DOI: 10.1615/PlasmaMed.2014011978. 54, 55

Hawtof R, Ghosh S, Guarr E, Xu C, Sankaran RM, Renner JN (2019) Catalyst–free, highly selective synthesis of ammonia from nitrogen and water by a plasma electrolytic system. *Science Advances* 5: eaat5778. DOI: 10.1126/sciadv.aat5778. 71, 77, 78

He C, Yu Y, Chen C, Yue L, Qiao N, Shen Q, Chen J, Hao Z (2013) Facile preparation of 3D ordered mesoporous CuO x–CeO 2 with notably enhanced efficiency for the low temperature oxidation of heteroatom–containing volatile organic compounds. *RSC Advances* 3: 19639–19656. DOI: 10.1039/c3ra42566e. 88

He J, Ma W, He J, Zhao J, Jimmy CY (2002) Photooxidation of azo dye in aqueous dispersions of H_2O_2/α–FeOOH. *Applied Catalysis B: Environmental* 39: 211–220. DOI: 10.1016/S0926-3373(02)00085-1. 86

Heijkers S, Aghaei M, Bogaerts A (2020) Plasma–Based CH4 Conversion into Higher Hydrocarbons and H2: Modeling to Reveal the Reaction Mechanisms of Different Plasma Sources. *The Journal of Physical Chemistry C* 124: 7016–7030. DOI: 10.1021/acs.jpcc.0c00082. 104

Heinlin J, Isbary G, Stolz W, Morfill G, Landthaler M, Shimizu T, Steffes B, Nosenko T, Zimmermann J, Karrer S (2011) Plasma applications in medicine with a special focus on dermatology. *Journal of the European Academy of Dermatology and Venereology* 25: 1–11. DOI: 10.1111/j.1468-3083.2010.03702.x. 40, 41

Heinlin J, Morfill G, Landthaler M, Stolz W, Isbary G, Zimmermann JL, Shimizu T, Karrer S (2010) *Plasma Medicine*: possible applications in dermatology. *JDDG: Journal der Deutschen Dermatologischen Gesellschaft* 8: 968–976. DOI: 10.1111/j.1610-0387.2010.07495.x. 40

Helmke A, Gerling T, Weltmann K–D (2018) Plasma Sources for Biomedical Applications. In: Comprehensive *Clinical Plasma Medicine*, pp. 23–41. Springer. DOI: 10.1007/978-3-319-67627-2_2. 4

Herron JT, Green DS (2001) Chemical kinetics database and predictive schemes for nonthermal humid air plasma chemistry. Part II. Neutral species reactions. *Plasma Chemistry and Plasma Processing* 21: 459–481. DOI: 10.1023/A:1011082611822. 15, 17, 18

Hertwig C, Leslie A, Meneses N, Reineke K, Rauh C, Schlüter O (2017) Inactivation of Salmonella Enteritidis PT30 on the surface of unpeeled almonds by cold plasma. *Innovative Food Science and Emerging Technologies* 44: 242–248. DOI: 10.1016/j.ifset.2017.02.007. 24, 37

Hertwig C, Meneses N, Mathys A (2018) Cold atmospheric pressure plasma and low energy electron beam as alternative nonthermal decontamination technologies for dry food surfaces: a review. *Trends in Food Science and Technology* 77: 131–142. DOI: 10.1016/j.tifs.2018.05.011. 31

Heuer K, Hoffmanns MA, Demir E, Baldus S, Volkmar CM, Röhle M, Fuchs PC, Awakowicz P, Suschek CV, Opländer C (2015) The topical use of non–thermal dielectric barrier discharge (DBD): Nitric oxide related effects on human skin. *Nitric Oxide* 44: 52–60. DOI: 10.1016/j.niox.2014.11.015. 41

Hoigné J, Bader H, Haag W, Staehelin J (1985) Rate constants of reactions of ozone with organic and inorganic compounds in water—III. Inorganic compounds and radicals. *Water Research* 19: 993–1004. DOI: 10.1016/0043-1354(85)90368-9. 19

Hojnik N, Cvelbar U, Tavčar–Kalcher G, Walsh JL, Križaj I (2017) Mycotoxin decontamination of food: cold atmospheric pressure plasma versus "classic" decontamination. *Toxins* 9: 151. DOI: 10.3390/toxins9050151. 31

Homola T, Pospíšil J, Krumpolec R, Souček P, Dzik P, Weiter M, Černák M (2018) Atmospheric dry hydrogen plasma reduction of inkjet-printed flexible graphene oxide electrodes. *ChemSusChem* 11: 941–947. DOI: 10.1002/cssc.201702139. 76

Hong J, Prawer S, Murphy AB (2018) Plasma catalysis as an alternative route for ammonia production: status, mechanisms, and prospects for progress. *ACS Sustainable Chemistry and Engineering* 6: 15–31. DOI: 10.1021/acssuschemeng.7b02381. 97, 105, 107

Hossain MM, Trinh QH, Nguyen DB, Sudhakaran M, Mok YS (2019) Robust hydrophobic coating on glass surface by an atmospheric–pressure plasma jet for plasma–polymerisation of hexamethyldisiloxane conjugated with (3–aminopropyl) triethoxysilane. *Surface Engineering* 35: 466–475. DOI: 10.1080/02670844.2018.1524037. 74

Houstis N, Rosen ED, Lander ES (2006) Reactive oxygen species have a causal role in multiple forms of insulin resistance. *Nature* 440: 944–948. DOI: 10.1038/nature04634. 41

Howland MF, Lele SK, Dabiri JO (2019) Wind farm power optimization through wake steering. *Proceedings of the National Academy of Sciences* 116: 14495–14500. DOI: 10.1073/pnas.1903680116. 95

Huang C, Chang Y–C, Wu S–Y (2010) Contact angle analysis of low–temperature cyclonic atmospheric pressure plasma modified polyethylene terephthalate. *Thin Solid Films* 518: 3575–3580. DOI: 10.1016/j.tsf.2009.11.046. 81

Huang C, Lin CC, Tsai CY, Juang RS (2013) Tailoring surface properties of polymeric separators for lithium-ion batteries by cyclonic atmospheric-pressure plasma. *Plasma Processes and Polymers* 10: 407–415. DOI: 10.1002/ppap.201200162. 98

Huang H, Ye D, Leung DY, Feng F, Guan X (2011) Byproducts and pathways of toluene destruction via plasma–catalysis. *Journal of Molecular Catalysis A: Chemical* 336: 87–93. DOI: 10.1016/j.molcata.2011.01.002. 91

Huang M, Zhuang H, Zhao J, Wang J, Yan W, Zhang J (2020) Differences in cellular damage induced by dielectric barrier discharge plasma between Salmonella Typhimurium and Staphylococcus aureus. *Bioelectrochemistry* 132: 107445. DOI: 10.1016/j.bioelechem.2019.107445. 35

Hübner S, Sousa JS, Van der Mullen J, Graham WG (2015) Thomson scattering on non–thermal atmospheric pressure plasma jets. *Plasma Sources Science and Technology* 24: 054005. DOI: 10.1088/0963-0252/24/5/054005. 9

Huie R, Padmaja S (1993) Free radical res commun 18: 195–199, pmid: 8396550. Vásquez–Vivar JM, Santos AM, Junqueira VBC, Augusto O (1996) *Biochem Journal* 314: 869–876. DOI: 10.3109/10715769309145868. 19

Irokawa Y, Morikawa T, Aoki K, Kosaka S, Ohwaki T, Taga Y (2006) Photodegradation of toluene over TiO2– X N X under visible light irradiation. *Physical Chemistry Chemical Physics* 8: 1116–1121. DOI: 10.1039/b517653k. 90

Isbary G, Heinlin J, Shimizu T, Zimmermann J, Morfill G, Schmidt HU, Monetti R, Steffes B, Bunk W, Li Y (2012) Successful and safe use of 2 min cold atmospheric argon plasma in chronic wounds: results of a randomized controlled trial. *British Journal of Dermatology* 167: 404–410. DOI: 10.1111/j.1365-2133.2012.10923.x. 39

Isbary G, Morfill G, Schmidt H, Georgi M, Ramrath K, Heinlin J, Karrer S, Landthaler M, Shimizu T, Steffes B (2010) A first prospective randomized controlled trial to decrease bacterial load using cold atmospheric argon plasma on chronic wounds in patients. *British Journal of Dermatology* 163: 78–82. DOI: 10.1111/j.1365-2133.2010.09744.x. 39, 40, 57

Isbary G, Morfill G, Zimmermann J, Shimizu T, Stolz W (2011) Cold atmospheric plasma: a successful treatment of lesions in Hailey–Hailey disease. *Archives of Dermatology* 147: 388–390. DOI: /10.1001/archdermatol.2011.57. 53

Isbary G, Shimizu T, Zimmermann J, Heinlin J, Al–Zaabi S, Rechfeld M, Morfill G, Karrer S, Stolz W (2014) Randomized placebo–controlled clinical trial showed cold atmospheric argon plasma relieved acute pain and accelerated healing in herpes zoster. *Clinical Plasma Medicine* 2: 50–55. DOI: 10.1016/j.cpme.2014.07.001. 41

Ischiropoulos H, Zhu L, Beckman JS (1992) Peroxynitrite formation from macrophage–derived nitric oxide. *Archives of Biochemistry and Biophysics* 298: 446–451. DOI: 10.1016/0003-9861(92)90433-W. 19

Iseki S, Nakamura K, Hayashi M, Tanaka H, Kondo H, Kajiyama H, Kano H, Kikkawa F, Hori M (2012) Selective killing of ovarian cancer cells through induction of apoptosis by nonequilibrium atmospheric pressure plasma. *Applied Physics Letters* 100: 113702. DOI: 10.1063/1.3694928. 44, 45

Iseni S, Schmidt–Bleker A, Winter J, Weltmann K–D, Reuter S (2014) Atmospheric pressure streamer follows the turbulent argon air boundary in a MHz argon plasma jet investigated by OH–tracer PLIF spectroscopy. *Journal of Physics D: Applied Physics* 47: 152001. DOI: 10.1088/0022-3727/47/15/152001. 13

Itikawa Y, Hayashi M, Ichimura A, Onda K, Sakimoto K, Takayanagi K, Nakamura M, Nishimura H, Takayanagi T (1986) Cross sections for collisions of electrons and photons with nitrogen molecules. *Journal of Physical and Chemical Reference Data* 15: 985–1010. DOI: 10.1063/1.555762. 15, 18

Ito M, Oh JS, Ohta T, Shiratani M, Hori M (2018) Current status and future prospects of agricultural applications using atmospheric-pressure plasma technologies. *Plasma Processes and Polymers* 15: 1700073. DOI: 10.1002/ppap.201700073. 23, 28, 29

Iwamoto M, Akiyama M, Aihara K, Deguchi T (2017) Ammonia synthesis on wool–like Au, Pt, Pd, Ag, or Cu electrode catalysts in nonthermal atmospheric–pressure plasma of N_2 and H_2. *ACS Catalysis* 7: 6924–6929. DOI: 10.1021/acscatal.7b01624. 87, 88

Iza F, Kim GJ, Lee SM, Lee JK, Walsh JL, Zhang YT, Kong MG (2008) Microplasmas: sources, particle kinetics, and biomedical applications. *Plasma Processes and Polymers* 5: 322–344. DOI: 10.1002/ppap.200700162. 35

Jablonowski H, von Woedtke T (2015) Research on *Plasma Medicine*–relevant plasma–liquid interaction: What happened in the past five years? *Clinical Plasma Medicine* 3: 42–52. DOI: 10.1016/j.cpme.2015.11.003. 28

Jang B, Helleson M, Shi C, Rondinone A, Schwartz V, Liang C, Overbury S (2008) Characterization of Al_2O_3 supported nickel catalysts derived from rf non–thermal plasma technology. *Topics in Catalysis* 49: 145–152. DOI: 10.1007/s11244-008-9091-2. 86

Jang BWL, Hammer T, Liu C–j (2004) Editorial. *Catalysis Today* 89: 1–2. DOI: 10.1016/S0920-5861(03)00560-1. 83

Jermann C, Koutchma T, Margas E, Leadley C, Ros–Polski V (2015) Mapping trends in novel and emerging food processing technologies around the world. *Innovative Food Science and Emerging Technologies* 31: 14–27. DOI: 10.1016/j.ifset.2015.06.007. 23

Ji S–H, Choi K–H, Pengkit A, Im JS, Kim JS, Kim YH, Park Y, Hong EJ, kyung Jung S, Choi E–H (2016) Effects of high voltage nanosecond pulsed plasma and micro DBD plasma on seed germination, growth development and physiological activities in spinach. *Archives of Biochemistry and Biophysics* 605: 117–128. DOI: 10.1016/j.abb.2016.02.028. 24

Jiafeng J, Xin H, Ling L, Jiangang L, Hanliang S, Qilai X, Renhong Y, Yuanhua D (2014) Effect of cold plasma treatment on seed germination and growth of wheat. *Plasma Science and Technology* 16: 54. 24

Jiang B, Zheng J, Qiu S, Wu M, Zhang Q, Yan Z, Xue Q (2014a) Review on electrical discharge plasma technology for wastewater remediation. *Chemical Engineering Journal* 236: 348–368. DOI: 10.1016/j.cej.2013.09.090. 63, 65

Jiang C, Zhang P, Zhang B, Li J, Wang M (2013a) Facile synthesis of activated carbon–supported porous manganese oxide via in situ reduction of permanganate for ozone decomposition. *Ozone: Science and Engineering* 35: 308–315. DOI: 10.1080/01919512.2013.795854. 90

Jiang J, Lu Y, Li J, Li L, He X, Shao H, Dong Y (2014b) Effect of seed treatment by cold plasma on the resistance of tomato to Ralstonia solanacearum (bacterial wilt). *PloS One* 9: e97753. DOI: 10.1371/journal.pone.0097753. 24

Jiang N, Hu J, Li J, Shang K, Lu N, Wu Y (2016) Plasma–catalytic degradation of benzene over Ag–Ce bimetallic oxide catalysts using hybrid surface/packed–bed discharge plasmas. *Applied Catalysis B: Environmental* 184: 355–363. DOI: 10.1016/j.apcatb.2015.11.044. 89, 92

Jiang N, Lu N, Shang K, Li J, Wu Y (2013b) Innovative approach for benzene degradation using hybrid surface/packed–bed discharge plasmas. *Environmental Science and Technology* 47: 9898–9903. DOI: 10.1021/es401110h. 61

Jin SY, Manuel J, Zhao X, Park WH, Ahn J–H (2017) Surface–modified polyethylene separator via oxygen plasma treatment for lithium ion battery. *Journal of Industrial and Engineering Chemistry* 45: 15–21. DOI: 10.1016/j.jiec.2016.08.021. 98

Johnson DL, Ambrose SH, Bassett TJ, Bowen ML, Crummey DE, Isaacson JS, Johnson DN, Lamb P, Saul M, Winter-Nelson AE (1997) Meanings of environmental terms. *Journal of Environmental Quality* 26: 581–589. DOI: 10.2134/jeq1997.00472425002600030002x. 59

Jolibois J, Takashima K, Mizuno A (2012) Application of a non–thermal surface plasma discharge in wet condition for gas exhaust treatment: NOx removal. *Journal of Electrostatics* 70: 300–308. DOI: 10.1016/j.elstat.2012.03.011. 62

Judée F, Dufour T (2019) Plasma gun for medical applications: engineering an equivalent electrical target of the human body and deciphering relevant electrical parameters. *Journal of Physics D: Applied Physics* 52: 16LT02. DOI: 10.1088/1361-6463/ab03b8. 35

Judée F, Simon S, Bailly C, Dufour T (2018) Plasma–activation of tap water using DBD for agronomy applications: Identification and quantification of long lifetime chemical species and production/consumption mechanisms. *Water Research* 133: 47–59. DOI: 10.1016/j.watres.2017.12.035. 28

Jung J, Han S, Kim B (2019) Digital numerical map–oriented estimation of solar energy potential for site selection of photovoltaic solar panels on national highway slopes. *Applied Energy* 242: 57–68. DOI: 10.1016/j.apenergy.2019.03.101. 95

Jurov A, Popović D, Rakić IŠ, Marion ID, Filipič G, Kovač J, Cvelbar U, Krstulović N (2019) Atmospheric pressure plasma jet–assisted impregnation of gold nanoparticles into PVC polymer for various applications. *The International Journal of Advanced Manufacturing Technology* 101: 927–938. DOI: 10.1007/s00170-018-2988-4. 71, 79, 80

Jwa E, Lee S, Lee H, Mok Y (2013) Plasma–assisted catalytic methanation of CO and CO2 over Ni–zeolite catalysts. *Fuel Processing Technology* 108: 89–93. DOI: 10.1016/j. fuproc.2012.03.008. 90

Kabouzi Y, Calzada M, Moisan M, Tran K, Trassy C (2002) Radial contraction of microwave–sustained plasma columns at atmospheric pressure. *Journal of Applied Physics* 91: 1008–1019. DOI: 10.1063/1.1425078. 8

Kaemling C, Kaemling A, Tümmel S, Viöl W (2005) Plasma treatment on finger nails prior to coating with a varnish. *Surface and Coatings Technology* 200: 668–671. DOI: 10.1016/j. surfcoat.2005.01.065. 53

Kalra CS, Kossitsyn M, Iskenderova K, Chirokov A, Cho YI, Gutsol A, Fridman A, (2003) Electrical discharges in the reverse vortex flow–tornado discharges, *Electronic Proceedings of 16th International Symposium on Plasma Chemistry*, Taormina, Italy. 4

Kamgang-Youbi G, Herry JM, Meylheuc T, Laminsi S, Naïtali M (2018) Microbial decontamination of stainless steel and polyethylene surfaces using GlidArc plasma activated water without chemical additives. *Journal of Chemical Technology and Biotechnology* 93: 2544–2551. DOI: 10.1002/jctb.5608. 28

Kanazawa S, Kogoma M, Moriwaki T, Okazaki S (1988) Stable glow plasma at atmospheric pressure. *Journal of Physics D: Applied Physics* 21: 838. DOI: 10.1088/0022-3727/21/5/028. 3

Karatum O, Deshusses MA (2016) A comparative study of dilute VOCs treatment in a non–thermal plasma reactor. *Chemical Engineering Journal* 294: 308–315. DOI: 10.1016/j. cej.2016.03.002. 60, 61, 84

Karuppiah J, Karvembu R, Subrahmanyam C (2012a) The catalytic effect of MnOx and CoOx on the decomposition of nitrobenzene in a non–thermal plasma reactor. *Chemical Engineering Journal* 180: 39–45. DOI: 10.1016/j.cej.2011.10.096. 92

Karuppiah J, Reddy EL, Reddy PMK, Ramaraju B, Karvembu R, Subrahmanyam C (2012b) Abatement of mixture of volatile organic compounds (VOCs) in a catalytic non–thermal plasma reactor. *Journal of Hazardous Materials* 237: 283–289. DOI: 10.1016/j. jhazmat.2012.08.040. 62

Katsumura Y (1998) NO_2 and NO_3 Radicals in radiolysis of nitric acid solutions. N–centered radicals. In: *The Chemistry of Free Radicals: N-Centered Radicals*. John Wiley & Sons. 19

Kauffman GB (1999) From triads to catalysis: Johann Wolfgang Döbereiner (1780–1849) on the 150th anniversary of his death. *The Chemical Educator* 4: 186–197. DOI: 10.1007/ s00897990326a. 83

Kaushik NK, Ghimire B, Li Y, Adhikari M, Veerana M, Kaushik N, Jha N, Adhikari B, Lee S–J, Masur K (2018) Biological and medical applications of plasma–activated media, water and solutions. *Biological Chemistry* 400: 39–62. DOI: 10.1515/hsz-2018-0226. 28

Kaushik NK, Kaushik N, Linh NN, Ghimire B, Pengkit A, Sornsakdanuphap J, Lee S–J, Choi EH (2019) Plasma and nanomaterials: fabrication and biomedical applications. *Nanomaterials* 9: 98. DOI: 10.3390/nano9010098. 22

Kazak O, Eker YR, Bingol H, Tor A (2017) Novel preparation of activated carbon by cold oxygen plasma treatment combined with pyrolysis. *Chemical Engineering Journal* 325: 564–575. DOI: 10.1016/j.cej.2017.05.107. 98

Kehrer M, Duchoslav J, Hinterreiter A, Mehic A, Stehrer T, Stifter D (2020) Surface functionalization of polypropylene using a cold atmospheric pressure plasma jet with gas water mixtures. *Surface and Coatings Technology* 384: 125170. DOI: 10.1016/j.surf-coat.2019.125170. 71

Keidar M (2015) Plasma for cancer treatment. *Plasma Sources Science and Technology* 24: 033001. DOI: 10.1088/0963-0252/24/3/033001. 22, 44

Keidar M (2018) A prospectus on innovations in the plasma treatment of cancer. *Physics of Plasmas* 25: 083504. DOI: 10.1063/1.5034355. 35, 43

Keidar M, Walk R, Shashurin A, Srinivasan P, Sandler A, Dasgupta S, Ravi R, Guerrero–Preston R, Trink B (2011) Cold plasma selectivity and the possibility of a paradigm shift in cancer therapy. *British Journal of Cancer* 105: 1295–1301. DOI: 10.1038/bjc.2011.386. 44, 45, 47, 48

Kelly S, Turner MM (2014) Generation of reactive species by an atmospheric pressure plasma jet. *Plasma Sources Science and Technology* 23: 065013. DOI: 10.1088/0963-0252/23/6/065013. 14, 15

Kerr JF, Wyllie AH, Currie AR (1972) Apoptosis: a basic biological phenomenon with wideranging implications in tissue kinetics. *British Journal of Cancer* 26: 239–257. DOI: 10.1038/bjc.1972.33. 35

Keyaerts E, Vijgen L, Chen L, Maes P, Hedenstierna G, Van Ranst M (2004) Inhibition of SARS–coronavirus infection *in vitro* by S–nitroso–N–acetylpenicillamine, a nitric oxide donor compound. *International Journal of Infectious Diseases* 8: 223–226. DOI: 10.1016/j.ijid.2004.04.012. 55

Khalili M, Daniels L, Lin A, Krebs FC, Snook AE, Bekeschus S, Bowne WB, Miller V (2019) Non–thermal plasma–induced immunogenic cell death in cancer. *Journal of Physics D: Applied Physics* 52: 423001. DOI: 10.1088/1361-6463/ab31c1. 53

Khamsen N, Onwimol D, Teerakawanich N, Dechanupaprittha S, Kanokbannakorn W, Hongesombut K, Srisonphan S (2016) Rice (Oryza sativa L.) seed sterilization and germination enhancement via atmospheric hybrid nonthermal discharge plasma. *ACS Applied Materials and Interfaces* 8: 19268–19275. DOI: 10.1021/acsami.6b04555. 26, 27

Khani MR, Pour EB, Rashnoo S, Tu X, Ghobadian B, Shokri B, Khadem A, Hosseini SI (2020) Real diesel engine exhaust emission control: indirect non–thermal plasma and comparison to direct plasma for NO_X, THC, CO, and CO_2. *Journal of Environmental Health Science and Engineering* 1–12. DOI: 10.1007/s40201-020-00500-0. 69

Khlyustova A, Labay C, Machala Z, Ginebra M–P, Canal C (2019a) Important parameters in plasma jets for the production of RONS in liquids for plasma medicine: A brief review. *Frontiers of Chemical Science and Engineering* 13: 238–252. DOI: 10.1007/s11705-019-1801-8. 5, 28

Khlyustova A, Sirotkin N, Kochkina N, Krayev A, Titov V, Agafonov A (2019b) Deposition of silver nanostructures on polymer films by glow discharge. *Plasma Chemistry and Plasma Processing* 39: 311–323. DOI: 10.1007/s11090-018-9932-0. 71

Kim H–H, Teramoto Y, Negishi N, Ogata A (2015a) A multidisciplinary approach to understand the interactions of nonthermal plasma and catalyst: A review. *Catalysis Today* 256: 13–22. DOI: 10.1016/j.cattod.2015.04.009. 90, 92, 93

Kim H–H, Teramoto Y, Ogata A, Takagi H, Nanba T (2016a) Plasma catalysis for environmental treatment and energy applications. *Plasma Chemistry and Plasma Processing* 36: 45–72. DOI: 10.1007/s11090-015-9652-7. 59, 62

Kim H–H, Teramoto Y, Sano T, Negishi N, Ogata A (2015b) Effects of Si/Al ratio on the interaction of nonthermal plasma and Ag/HY catalysts. *Applied Catalysis B: Environmental* 166: 9–17. DOI: 10.1016/j.apcatb.2014.11.008. 91

Kim H–J, Won C–H, Kim H–W (2018) Pathogen deactivation of glow discharge cold plasma while treating organic and inorganic pollutants of slaughterhouse wastewater. *Water, Air, and Soil Pollution* 229: 237. DOI: 10.1007/s11270-018-3895-x. 103

Kim HH (2004) Nonthermal plasma processing for air-pollution control: a historical review, current issues, and future prospects. *Plasma Processes and Polymers* 1: 91–110. DOI: 10.1002/ppap.200400028. 62, 107

Kim HH, Teramoto Y, Ogata A, Takagi H, Nanba T (2017) Atmospheric-pressure nonthermal plasma synthesis of ammonia over ruthenium catalysts. *Plasma Processes and Polymers* 14: 1600157. DOI: 10.1002/ppap.201600157. 71, 77

Kim HY, Kang SK, Park SM, Jung HY, Choi BH, Sim JY, Lee JK (2015c) Characterization and effects of Ar/Air microwave plasma on wound healing. *Plasma Processes and Polymers* 12: 1423–1434. DOI: 10.1002/ppap.201500017. 41, 43

Kim J, Kim JH, Chang B, Choi EH, Park H–K (2016b) Hemorheological alterations of red blood cells induced by non–thermal dielectric barrier discharge plasma. *Applied Physics Letters* 109: 193701. DOI: 10.1063/1.4967451. 53

Kim J–W, Puligundla P, Mok C (2015d) Microbial decontamination of dried laver using corona discharge plasma jet (CDPJ). *Journal of Food Engineering* 161: 24–32. DOI: 10.1016/j.jfoodeng.2015.03.034. 37

Kim K–R, Owens G, Kwon S–I, So K–H, Lee D–B, Ok YS (2011) Occurrence and environmental fate of veterinary antibiotics in the terrestrial environment. *Water, Air, and Soil Pollution* 214: 163–174. DOI: 10.1007/s11270-010-0412-2. 66

Kim K–S, Yang C–S, Mok Y (2013) Degradation of veterinary antibiotics by dielectric barrier discharge plasma. *Chemical Engineering Journal* 219: 19–27. DOI: 10.1016/j.cej.2012.12.079. 66, 71

Kim SY, Bang IH, Min SC (2019) Effects of packaging parameters on the inactivation of Salmonella contaminating mixed vegetables in plastic packages using atmospheric dielectric barrier discharge cold plasma treatment. *Journal of Food Engineering* 242: 55–67. DOI: 10.1016/j.jfoodeng.2018.08.020. 30

Kim Y–J, Jin S, Han G–H, Kwon GC, Choi JJ, Choi EH, Uhm HS, Cho G (2015e) Plasma apparatuses for biomedical applications. *IEEE Transactions on Plasma Science* 43: 944–950. DOI: 10.1109/TPS.2015.2388775. 7

Klockow PA, Keener KM (2009) Safety and quality assessment of packaged spinach treated with a novel ozone–generation system. *LWT–Food Science and Technology* 42: 1047–1053. DOI: 10.1016/j.lwt.2009.02.011. 31

Knorr D, Froehling A, Jaeger H, Reineke K, Schlueter O, Schoessler K (2011) Emerging technologies in food processing. *Annual Review of Food Science and Technology* 2: 203–235. DOI: 10.1146/annurev.food.102308.124129. 23

Koga K, Thapanut S, Amano T, Seo H, Itagaki N, Hayashi N, Shiratani M (2015) Simple method of improving harvest by nonthermal air plasma irradiation of seeds of Arabidopsis thaliana (L.). *Applied Physics Express* 9: 016201. DOI: 10.7567/APEX.9.016201. 24, 28

Kogelschatz U (2003) Dielectric–barrier discharges: their history, discharge physics, and industrial applications. *Plasma Chemistry and Plasma Processing* 23: 1–46. DOI: 10.1023/A:1022470901385. 8

Kong MG, Kroesen G, Morfill G, Nosenko T, Shimizu T, Van Dijk J, Zimmermann J (2009) *Plasma Medicine*: an introductory review. *New Journal of Physics* 11: 115012. DOI: 10.1088/1367-2630/11/11/115012. 5

Kong R, Jia G, Cheng Z–x, Wang Y–w, Mu M, Wang S–j, Pan S–h, Gao Y, Jiang H–c, Dong D–l (2012) Dihydroartemisinin enhances Apo2L/TRAIL–mediated apoptosis in pancreatic cancer cells via ROS–mediated up–regulation of death receptor 5. *PloS One* 7: e37222. DOI: 10.1371/journal.pone.0037222. 45

Kortshagen UR, Sankaran RM, Pereira RN, Girshick SL, Wu JJ, Aydil ES (2016) Nonthermal plasma synthesis of nanocrystals: fundamental principles, materials, and applications. *Chemical Reviews* 116: 11061–11127. DOI: 10.1021/acs.chemrev.6b00039. 5, 80, 105

Kossyi I, Kostinsky AY, Matveyev A, Silakov V (1992) Kinetic scheme of the non–equilibrium discharge in nitrogen–oxygen mixtures. *Plasma Sources Science and Technology* 1: 207. DOI: 10.1088/0963-0252/1/3/011. 14, 15

Kostov KG, Nishime TMC, Castro AHR, Toth A, Hein LRdO (2014) Surface modification of polymeric materials by cold atmospheric plasma jet. *Applied Surface Science* 314: 367–375. DOI: 10.1016/j.apsusc.2014.07.009. 71, 72, 103

Kouzarides T (2007) Chromatin modifications and their function. *Cell* 128: 693–705. DOI: 10.1016/j.cell.2007.02.005. 54

Kramer A, Schauer F, Papke R, Bekeschus S (2018) Plasma application for hygienic purposes in medicine, industry, and biotechnology: Update 2017. In: *Comprehensive Clinical Plasma Medicine*, pp. 253–281. Springer. DOI: 10.1007/978-3-319-67627-2_14. 75

Kramer N, Aydil E, Kortshagen U (2015) Requirements for plasma synthesis of nanocrystals at atmospheric pressures. *Journal of Physics D: Applied Physics* 48: 035205. DOI: 10.1088/0022-3727/48/3/035205. 80

Krishnamurthy A, Adebayo B, Gelles T, Rownaghi A, Rezaei F (2019) Abatement of gaseous volatile organic compounds: A process perspective. *Catalysis Today*. DOI: 10.1016/j.cattod.2019.05.069. 90

Kubinova S, Zaviskova K, Uherkova L, Zablotskii V, Churpita O, Lunov O, Dejneka A (2017) Non–thermal air plasma promotes the healing of acute skin wounds in rats. *Scientific Reports* 7: 1–11. DOI: 10.1038/srep45183. 41

Kusano Y (2014) Atmospheric pressure plasma processing for polymer adhesion: a review. *The Journal of Adhesion* 90: 755–777. DOI: /10.1080/00218464.2013.804407. 74

Kuzminova A, Kretková T, Kylián O, Hanuš J, Khalakhan I, Prukner V, Doležalová E, Šimek M, Biederman H (2017) Etching of polymers, proteins and bacterial spores by atmospheric pressure DBD plasma in air. *Journal of Physics D: Applied Physics* 50: 135201. DOI: 10.1088/1361-6463/aa5c21. 73, 75

Kyrychenko A, Pasko DA, Kalugin ON (2017) Poly (vinyl alcohol) as a water protecting agent for silver nanoparticles: the role of polymer size and structure. *Physical Chemistry Chemical Physics* 19: 8742–8756. DOI: 10.1039/C6CP05562A. 77

Lang C, Ouyang L, Yang L, Dai L, Wu D, Shao H, Zhu M (2018) Enhanced hydrogen storage kinetics in Mg@ FLG composite synthesized by plasma assisted milling. *International Journal of Hydrogen Energy* 43: 17346–17352. DOI: 10.1016/j.ijhydene.2018.07.149. 71, 97

Langmuir I (1928) Oscillations in ionized gases. *Proceedings of the National Academy of Sciences of the United States of America* 14: 627. DOI: 10.1073/pnas.14.8.627. 1

Larcher D, Tarascon J–M (2015) Toward greener and more sustainable batteries for electrical energy storage. *Nature Chemistry* 7: 19–29. DOI: 10.1038/nchem.2085. 95

Laroussi M (1996) Sterilization of contaminated matter with an atmospheric pressure plasma. *IEEE Transactions on Plasma Science* 24: 1188–1191. DOI: 10.1109/27.533129. 35, 36

Laroussi M (2002) Nonthermal decontamination of biological media by atmospheric–pressure plasmas: review, analysis, and prospects. *IEEE Transactions on Plasma Science* 30: 1409–1415. DOI: 10.1109/TPS.2002.804220. 37

Laroussi M (2018) Plasma Medicine: a brief introduction. *Plasma* 1: 47–60. DOI: 10.3390/plasma1010005. 35

Laroussi M (2019) 1995–2005: A decade of innovation in low temperature plasma and its applications. *Plasma* 2: 360–368. DOI: 10.3390/plasma2030028. 35

Laroussi M, Kong M, Morfill G (2012) *Plasma Medicine: Applications of Low–temperature Gas Plasmas in Medicine and Biology.* Cambridge University Press. DOI: 10.1017/CBO9780511902598. 104

Laroussi M, Leipold F (2004) Evaluation of the roles of reactive species, heat, and UV radiation in the inactivation of bacterial cells by air plasmas at atmospheric pressure. *International Journal of Mass Spectrometry* 233: 81–86. DOI: 10.1016/j.ijms.2003.11.016. 37

Laroussi M, Lu X (2005) Room–temperature atmospheric pressure plasma plume for biomedical applications. *Applied Physics Letters* 87: 113902. DOI: 10.1063/1.2045549. 9

Laroussi M, Lu X, Keidar M (2017) Perspective: The physics, diagnostics, and applications of atmospheric pressure low temperature plasma sources used in *Plasma Medicine. Journal of Applied Physics* 122: 020901. DOI: 10.1063/1.4993710. 7, 22

Lau T, Chan LK, Wong EC, Hui CW, Sneddon K, Cheung T, Yim S, Lee JH, Yeung CS, Chung TK (2017) A loop of cancer–stroma–cancer interaction promotes peritoneal metastasis of ovarian cancer via TNFα–TGFα–EGFR. *Oncogene* 36: 3576–3587. DOI: 10.1038/onc.2016.509. 52

Le Pape H, Solano–Serena F, Contini P, Devillers C, Maftah A, Leprat P (2004) Involvement of reactive oxygen species in the bactericidal activity of activated carbon fibre supporting silver: Bactericidal activity of ACF (Ag) mediated by ROS. *Journal of Inorganic Biochemistry* 98: 1054–1060. DOI: 10.1016/j.jinorgbio.2004.02.025. 77

Lee B–J, Jeong G–H (2018) Efficient surface functionalization of vertically–aligned carbon nanotube arrays using an atmospheric pressure plasma jet system. *Fullerenes, Nanotubes, and Carbon Nanostructures* 26: 116–122. DOI: 10.1080/1536383X.2017.1409211. 71, 74

Lee H, Lee D–H, Song Y–H, Choi WC, Park Y–K, Kim DH (2015) Synergistic effect of non–thermal plasma–catalysis hybrid system on methane complete oxidation over Pd–based catalysts. *Chemical Engineering Journal* 259: 761–770. DOI: 10.1016/j.cej.2014.07.128. 89, 90, 92

Lee H, Park G, Seo Y, Im Y, Shim S, Lee H (2011) Modelling of atmospheric pressure plasmas for biomedical applications. *Journal of Physics D: Applied Physics* 44: 053001. DOI: 10.1088/0022-3727/44/5/053001. 7, 13

Lee HM, Chang MB (2003) Abatement of gas–phase p–xylene via dielectric barrier discharges. *Plasma Chemistry and Plasma Processing* 23: 541–558. DOI: 10.1023/A:1023239122885. 90

Lei C, Su B, Dong H, Bellavia A, Di Fenza R, Fakhr BS, Gianni S, Grassi LG, Kacmarek R, Morais CCA (2020) Protocol of a randomized controlled trial testing inhaled Nitric Oxide in mechanically ventilated patients with severe acute respiratory syndrome in COVID–19 (SARS–CoV–2). *medRxiv*. DOI: 10.1101/2020.03.09.20033530. 55

Levchenko I, Bazaka K, Baranov O, Sankaran RM, Nomine A, Belmonte T, Xu S (2018) Lightning under water: Diverse reactive environments and evidence of synergistic effects for material treatment and activation. *Applied Physics Reviews* 5: 021103. DOI: 10.1063/1.5024865. 103

Levchenko I, Ostrikov KK, Zheng J, Li X, Keidar M, Teo KB (2016) Scalable graphene production: perspectives and challenges of plasma applications. *Nanoscale* 8: 10511–10527. DOI: 10.1039/C5NR06537B. 80

Ley H–H (2014) Analytical methods in plasma diagnostic by optical emission spectroscopy: A tutorial review. *Journal of Science and Technology* 6. 12

Li J, Chen C, Wei J, Li J, Wang X (2014a) Enhanced electrochemical performance of reduced graphene oxides by H2/Ar plasma treatment. *The Journal of Physical Chemistry* C 118: 28440–28447. DOI: 10.1021/jp509182g. 76

Li J, Ma C, Zhu S, Yu F, Dai B, Yang D (2019a) A review of recent advances of dielectric barrier discharge plasma in catalysis. *Nanomaterials* 9: 1428. DOI: 10.3390/nano9101428. 105

Li S–W, Xue L (2010) The interaction between H_2O_2 and NO, Ca_2^+, cGMP, and MAPKs during adventitious rooting in mung bean seedlings. *In Vitro Cellular and Developmental Biology– Plant* 46: 142–148. DOI: 10.1007/s11627-009-9275-x. 29

Li W, Liu X, Khan MA, Yamaguchi S (2005) The effect of plant growth regulators, nitric oxide, nitrate, nitrite and light on the germination of dimorphic seeds of Suaeda salsa under saline conditions. *Journal of Plant Research* 118: 207–214. DOI: 10.1007/s10265-005-0212-8. 24

Li W, Yu H, Ding D, Chen Z, Wang Y, Wang S, Li X, Keidar M, Zhang W (2019b) Cold atmospheric plasma and iron oxide–based magnetic nanoparticles for synergetic lung cancer therapy. *Free Radical Biology and Medicine* 130: 71–81. DOI: 10.1016/j.freeradbiomed.2018.10.429. 44, 51

Li X, Farid M (2016) A review on recent development in non–conventional food sterilization technologies. *Journal of Food Engineering* 182: 33–45. DOI: 10.1016/j.jfoodeng.2016.02.026. 23

Li X, Zhang H, Luo Y, Teng Y (2014b) Remediation of soil heavily polluted with polychlorinated biphenyls using a low–temperature plasma technique. *Frontiers of Environmental Science and Engineering* 8: 277–283. DOI: 10.1007/s11783-013-0562-8. 67

Li Y, Fan Z, Shi J, Liu Z, Zhou J, Shangguan W (2015) Modified manganese oxide octahedral molecular sieves M′–OMS–2 (M′= Co, Ce, Cu) as catalysts in post plasma–catalysis for acetaldehyde degradation. *Catalysis Today* 256: 178–185. DOI: 10.1016/j.cattod.2015.02.003. 89

Li Y, Jang BW (2011) Non–thermal RF plasma effects on surface properties of Pd/TiO2 catalysts for selective hydrogenation of acetylene. *Applied Catalysis A: General* 392: 173–179. DOI: 10.1016/j.apcata.2010.11.008. 84

Liang Y, Wu Y, Sun K, Chen Q, Shen F, Zhang J, Yao M, Zhu T, Fang J (2012) Rapid inactivation of biological species in the air using atmospheric pressure nonthermal plasma. *Environmental Science and Technology* 46: 3360–3368. DOI: 10.1021/es203770q. 63

Liao X, Cullen P, Muhammad AI, Jiang Z, Ye X, Liu D, Ding T (2020) Cold plasma–based hurdle interventions: New strategies for improving food safety. *Food Engineering Reviews* 1–12. DOI: 10.1007/s12393-020-09222-3. 103

Liao X, Liu D, Chen S, Ye X, Ding T (2019) Degradation of antibiotic resistance contaminants in wastewater by atmospheric cold plasma: kinetics and mechanisms. *Environmental Technology* 1–14. DOI: 10.1080/09593330.2019.1620866. 107

Liao X, Liu D, Xiang Q, Ahn J, Chen S, Ye X, Ding T (2017) Inactivation mechanisms of non–thermal plasma on microbes: A review. *Food Control* 75: 83–91. DOI: 10.1016/j.foodcont.2016.12.021. 76

Lieberman MA, Lichtenberg AJ (2005) *Principles of Plasma Discharges and Materials Processing.* John Wiley and Sons. DOI: 10.1002/0471724254. 71

Liedtke KR, Bekeschus S, Kaeding A, Hackbarth C, Kuehn J–P, Heidecke C–D, von Bernstorff W, von Woedtke T, Partecke LI (2017) Non–thermal plasma–treated solution demonstrates antitumor activity against pancreatic cancer cells *in vitro* and *in vivo. Scientific Reports* 7: 1–12. DOI: 10.1038/s41598-017-08560-3. 46

Liguori A, Gallingani T, Padmanaban DB, Laurita R, Velusamy T, Jain G, Macias–Montero M, Mariotti D, Gherardi M (2018) Synthesis of copper–based nanostructures in liquid environments by means of a non–equilibrium atmospheric pressure nanopulsed plasma jet. *Plasma Chemistry and Plasma Processing* 38: 1209–1222. DOI: 10.1007/s11090-018-9924-0. 79, 80

Lin A, Truong B, Patel S, Kaushik N, Choi EH, Fridman G, Fridman A, Miller V (2017) Nano-second–pulsed DBD plasma–generated reactive oxygen species trigger immunogenic cell death in A549 lung carcinoma cells through intracellular oxidative stress. *International Journal of Molecular Sciences* 18: 966. DOI: 10.3390/ijms18050966. 53

Lin AG, Xiang B, Merlino DJ, Baybutt TR, Sahu J, Fridman A, Snook AE, Miller V (2018a) Non–thermal plasma induces immunogenic cell death *in vivo* in murine CT26 colorectal tumors. *Oncoimmunology* 7: e1484978. DOI: 10.1080/2162402X.2018.1484978. 53

Lin HH, Yang H, Wu CY, Hsu SY, Duh J–G (2018b) Atmospheric pressure plasma jet irradiation of N doped carbon decorated on Li4Ti5O12 electrode as a high rate anode material for lithium ion batteries. *Journal of The Electrochemical Society* 166: A5146. DOI: 10.1149/2.0201903jes. 98

Lin ZQ, Kondo T, Ishida Y, Takayasu T, Mukaida N (2003) Essential involvement of IL-6 in the skin wound-healing process as evidenced by delayed wound healing in IL-6-deficient mice. *Journal of Leukocyte Biology* 73: 713–721. DOI: 10.1189/jlb.0802397. 39

Lindsay J, Degli Esposti M, Gilmore AP (2011) Bcl–2 proteins and mitochondria—specificity in membrane targeting for death. *Biochimica et Biophysica Acta (BBA)–Molecular Cell Research* 1813: 532–539. DOI: 10.1016/j.bbamcr.2010.10.017. 50

Ling L, Jiangang L, Minchong S, Chunlei Z, Yuanhua D (2015) Cold plasma treatment enhances oilseed rape seed germination under drought stress. *Scientific Reports* 5: 1–10. DOI: 10.1038/srep13033. 24

Linstrom PJ, Mallard WG (2001) The NIST Chemistry WebBook: A chemical data resource on the internet. *Journal of Chemical and Engineering Data* 46: 1059–1063. DOI: 10.1021/je000236i. 17

Lisperguer J, Perez P, Urizar S (2009) Structure and thermal properties of lignins: characterization by infrared spectroscopy and differential scanning calorimetry. *Journal of the Chilean Chemical Society* 54: 460–463. DOI: 10.4067/S0717-97072009000400030. 27

Liu C, Chen C, Jiang A, Sun X, Guan Q, Hu W (2020a) Effects of plasma–activated water on microbial growth and storage quality of fresh–cut apple. *Innovative Food Science and Emerging Technologies* 59: 102256. DOI: 10.1016/j.ifset.2019.102256. 28

Liu C–j, Vissokov GP, Jang BW–L (2002) Catalyst preparation using plasma technologies. *Catalysis Today* 72: 173–184. DOI: 10.1016/S0920-5861(01)00491-6. 83, 93

Liu D, Zhang Y, Xu M, Chen H, Lu X, Ostrikov K (2020b) Cold atmospheric pressure plasmas in dermatology: Sources, reactive agents, and therapeutic effects. *Plasma Processes and Polymers* 17: 1900218. DOI: 10.1002/ppap.201900218. 53

Liu F, Zhang B, Fang Z, Wan M, Wan H, Ostrikov K (2018) Jet-to-jet interactions in atmospheric-pressure plasma jet arrays for surface processing. *Plasma Processes and Polymers* 15: 1700114. DOI: 10.1002/ppap.201700114. 71, 72

Liu G, Yue R, Jia Y, Ni Y, Yang J, Liu H, Wang Z, Wu X, Chen Y (2013) Catalytic oxidation of benzene over Ce–Mn oxides synthesized by flame spray pyrolysis. *Particuology* 11: 454–459. DOI: 10.1016/j.partic.2012.09.013. 87

Liu J, Sonshine DA, Shervani S, Hurt RH (2010) Controlled release of biologically active silver from nanosilver surfaces. *ACS Nano* 4: 6903–6913. DOI: 10.1021/nn102272n. 77

Liu S–s (1997) Generating, partitioning, targeting and functioning of superoxide in mitochondria. *Bioscience Reports* 17: 259–272. DOI: 10.1023/A:1027328510931. 50

Liu X, Mou C–Y, Lee S, Li Y, Secrest J, Jang BW–L (2012) Room temperature O2 plasma treatment of SiO2 supported Au catalysts for selective hydrogenation of acetylene in the presence of large excess of ethylene. *Journal of Catalysis* 285: 152–159. DOI: 10.1016/j.jcat.2011.09.025. 85, 86

Liu X–L, Zhang X–T, Meng J, Zhang H–F, Zhao Y, Li C, Sun Y, Mei Q–B, Zhang F, Zhang T (2017) ING5 knockdown enhances migration and invasion of lung cancer cells by inducing EMT via EGFR/PI3K/Akt and IL–6/STAT3 signaling pathways. *Oncotarget* 8: 54265. DOI: 10.18632/oncotarget.17346. 52

Lloyd G, Friedman G, Jafri S, Schultz G, Fridman A, Harding K (2010) Gas plasma: medical uses and developments in wound care. *Plasma Processes and Polymers* 7: 194–211. DOI: 10.1002/ppap.200900097. 39, 40

Locke BR, Shih K–Y (2011) Review of the methods to form hydrogen peroxide in electrical discharge plasma with liquid water. *Plasma Sources Science and Technology* 20: 034006. DOI: 10.1088/0963-0252/20/3/034006. 19, 64

Los A, Ziuzina D, Akkermans S, Boehm D, Cullen PJ, Van Impe J, Bourke P (2018) Improving microbiological safety and quality characteristics of wheat and barley by high voltage atmospheric cold plasma closed processing. *Food Research International* 106: 509–521. DOI: 10.1016/j.foodres.2018.01.009. 24

Lou J, Lu N, Li J, Wang T, Wu Y (2012) Remediation of chloramphenicol–contaminated soil by atmospheric pressure dielectric barrier discharge. *Chemical Engineering Journal* 180: 99–105. DOI: 10.1016/j.cej.2011.11.013. 67

Loyer F, Bengasi G, Frache G, Choquet P, Boscher ND (2018a) Insights in the initiation and termination of poly (alkyl acrylates) synthesized by atmospheric pressure plasma-initiated chemical vapor deposition (AP-PiCVD). *Plasma Processes and Polymers* 15: 1800027. DOI: 10.1002/ppap.201800027. 71, 76

Loyer F, Bulou S, Choquet P, Boscher ND (2018b) Pulsed plasma initiated chemical vapor deposition (PiCVD) of polymer layers– A kinetic model for the description of gas phase to surface interactions in pulsed plasma discharges. *Plasma Processes and Polymers* 15: 1800121. DOI: 10.1002/ppap.201800121. 71, 76

Lu X, Keidar M, Laroussi M, Choi E, Szili E, Ostrikov K (2019) Transcutaneous plasma stress: From soft–matter models to living tissues. *Materials Science and Engineering: R: Reports* 138: 36–59. DOI: 10.1016/j.mser.2019.04.002. 6, 57, 104, 106

Lu X, Naidis G, Laroussi M, Ostrikov K (2014) Guided ionization waves: Theory and experiments. *Physics Reports* 540: 123–166. DOI: 10.1016/j.physrep.2014.02.006. 2

Lu X, Naidis G, Laroussi M, Reuter S, Graves D, Ostrikov K (2016) Reactive species in non–equilibrium atmospheric–pressure plasmas: Generation, transport, and biological effects. *Physics Reports* 630: 1–84. DOI: 10.1016/j.physrep.2016.03.003. 5, 9, 10, 21, 103, 106

Lu X, Ostrikov K (2018) Guided ionization waves: The physics of repeatability. *Applied Physics Reviews* 5: 031102. DOI: 10.1063/1.5031445. 2, 103

Luan P, Oehrlein G (2018) Stages of polymer transformation during remote plasma oxidation (RPO) at atmospheric pressure. *Journal of Physics D: Applied Physics* 51: 135201. DOI: 10.1088/1361-6463/aaaf60. 71, 89

Luk P, Sinha S, Lord R (2005) Upregulation of inducible nitric oxide synthase (iNOS) expression in faster–healing leg ulcers. *Journal of Wound Care* 14: 373–381. DOI: 10.12968/jowc.2005.14.8.26826. 41

Lukes P, Dolezalova E, Sisrova I, Clupek M (2014) Aqueous–phase chemistry and bactericidal effects from an air discharge plasma in contact with water: evidence for the formation of peroxynitrite through a pseudo–second–order post–discharge reaction of H_2O_2 and HNO_2. *Plasma Sources Science and Technology* 23: 015019. DOI: 10.1088/0963-0252/23/1/015019. 19, 28

Lukes P, Locke BR, Brisset J–L (2012) Aqueous–phase chemistry of electrical discharge plasma in water and in gas–liquid environments. *Plasma Chemistry and Catalysis in Gases and Liquids* 1: 243–308. DOI: 10.1002/9783527649525.ch7. 64

Luo J–d, Chen AF (2005) Nitric oxide: a newly discovered function on wound healing. *Acta Pharmacologica Sinica* 26: 259–264. DOI: 10.1111/j.1745-7254.2005.00058.x. 41

Ly L, Jones S, Shashurin A, Zhuang T, Rowe W, Cheng X, Wigh S, Naab T, Keidar M, Canady J (2018) A new cold plasma jet: Performance evaluation of cold plasma, hybrid plasma and argon plasma coagulation. *Plasma* 1: 189–200. DOI: 10.3390/plasma1010017. 53

Lymar SV, Schwarz HA, Czapski G (2002) Reactions of the Dihydroxylamine ($HNO_2^{-•}$) and Hydronitrite ($NO_2 2^{-•}$) Radical Ions. *The Journal of Physical Chemistry* A 106: 7245–7250. DOI: 10.1021/jp026107l. 19

Ma R, Wang G, Tian Y, Wang K, Zhang J, Fang J (2015) Non–thermal plasma–activated water inactivation of food–borne pathogen on fresh produce. *Journal of Hazardous Materials* 300: 643–651. DOI: /10.1016/j.jhazmat.2015.07.061. 31

Ma S, Zhao Y, Yang J, Zhang S, Zhang J, Zheng C (2017) Research progress of pollutants removal from coal–fired flue gas using non–thermal plasma. *Renewable and Sustainable Energy Reviews* 67: 791–810. DOI: 10.1016/j.rser.2016.09.066. 60

Ma X, Wu M, Liu S, Huang J, Sun B, Zhou Y, Zhu Q, Lu H (2019) Concentration control of volatile organic compounds by ionic liquid absorption and desorption. *Chinese Journal of Chemical Engineering* 27: 2383–2389. DOI: 10.1016/j.cjche.2018.12.019. 90

Macedo MJ, Silva GS, Feitor MC, Costa TH, Ito EN, Melo JD (2020) Surface modification of kapok fibers by cold plasma surface treatment. *Journal of Materials Research and Technology*. DOI: 10.1016/j.jmrt.2019.12.077. 104

Maguire G (2020) Stem cells, part of the innate and adaptive immune systems, as an antimicrobial for coronavirus Covid–19. *Stem Cells, Health, Technology.* 56

Magureanu M, Mandache N, Parvulescu V, Subrahmanyam C, Renken A, Kiwi–Minsker L (2007) Improved performance of non–thermal plasma reactor during decomposition of trichloroethylene: Optimization of the reactor geometry and introduction of catalytic electrode. *Applied Catalysis B: Environmental* 74: 270–277. DOI: 10.1016/j.apcatb.2007.02.019. 92

Magureanu M, Mandache NB, Parvulescu VI (2015) Degradation of pharmaceutical compounds in water by non–thermal plasma treatment. *Water Research* 81: 124–136. DOI: 10.1016/j.watres.2015.05.037. 66, 84

Magureanu M, Piroi D, Mandache N, David V, Medvedovici A, Bradu C, Parvulescu V (2011) Degradation of antibiotics in water by non–thermal plasma treatment. *Water Research* 45: 3407–3416. DOI: 10.1016/j.watres.2011.03.057. 64

Mahnot NK, Siyu L–P, Wan Z, Keener KM, Misra N (2020) In–package cold plasma decontamination of fresh–cut carrots: microbial and quality aspects. *Journal of Physics D: Applied Physics* 53: 154002. DOI: 10.1088/1361-6463/ab6cd3. 30

Malik MA, Ghaffar A, Malik SA (2001) Water purification by electrical discharges. *Plasma Sources Science and Technology* 10: 82. DOI: 10.1088/0963-0252/10/1/311. 64

Malik MA, Kolb JF, Sun Y, Schoenbach KH (2011) Comparative study of NO removal in surface–plasma and volume–plasma reactors based on pulsed corona discharges. *Journal of Hazardous Materials* 197: 220–228. DOI: 10.1016/j.jhazmat.2011.09.079. 60, 62

Mandal R, Singh A, Singh AP (2018) Recent developments in cold plasma decontamination technology in the food industry. *Trends in Food Science and Technology* 80: 93–103. DOI: 10.1016/j.tifs.2018.07.014. 23

Mangun CL, Benak KR, Economy J, Foster KL (2001) Surface chemistry, pore sizes and adsorption properties of activated carbon fibers and precursors treated with ammonia. *Carbon* 39: 1809–1820. DOI: 10.1016/S0008-6223(00)00319-5. 67

Manion J, Huie R, Levin R, Burgess Jr D, Orkin V, Tsang W, McGivern W, Hudgens J, Knyazev V, Atkinson D (2015) *NIST Chemical Kinetics Database, NIST Standard Reference Database 17*, Version 7.0 (Web Version), Release 1.6. 8, Data version 2015.12, National Institute of Standards and Technology, Gaithersburg, Maryland, 20899–8320. URL: http://kinetics nist gov (дата обращения: 2701 2018). 15, 16, 17, 18, 19

Mariotti D, Sankaran RM (2011) Perspectives on atmospheric–pressure plasmas for nano-fabrication. *Journal of Physics D: Applied Physics* 44: 174023. DOI: 10.1088/0022-3727/44/17/174023. 71

Marković M, Jović M, Stanković D, Kovačević V, Roglić G, Gojgić–Cvijović G, Manojlović D (2015) Application of non–thermal plasma reactor and Fenton reaction for degradation of ibuprofen. *Science of the Total Environment* 505: 1148–1155. DOI: /10.1016/j.scitotenv.2014.11.017. 64

Marsh K, Bugusu B (2007) Food packaging—roles, materials, and environmental issues. *Journal of Food Science* 72: R39–R55. DOI: 10.1111/j.1750-3841.2007.00301.x. 31

Martel J, Ko Y–F, Young JD, Ojcius DM (2020) Could nasal nitric oxide help to mitigate the severity of COVID–19? *Microbes and Infection* 22: 168–171. DOI: 10.1016/j.micinf.2020.05.002. 55

Mateu–Sanz M, Tornín J, Brulin B, Khlyustova A, Ginebra M–P, Layrolle P, Canal C (2020) Cold plasma–treated ringer's saline: A weapon to target osteosarcoma. *Cancers* 12: 227. DOI: 10.3390/cancers12010227. 35

Mattox DM, Mattox V (2003) *Vacuum Coating Technology*. Springer. 71

Medvecká V, Kováčik D, Zahoranová A, Černák M (2018) Atmospheric pressure plasma assisted calcination by the preparation of TiO_2 fibers in submicron scale. *Applied Surface Science* 428: 609–615. DOI: 10.1016/j.apsusc.2017.09.178. 71

Meemusaw M, Magaraphan R, (2013) Oxygen cold plasma treatment to improve hydrophilicity of HDPE pellets. *Advanced Materials Research. Trans Tech Publications*, pp. 210–213. DOI: 10.4028/www.scientific.net/AMR.747.210. 54

Mehta P, Barboun P, Herrera FA, Kim J, Rumbach P, Go DB, Hicks JC, Schneider WF (2018) Overcoming ammonia synthesis scaling relations with plasma–enabled catalysis. *Nature Catalysis* 1: 269–275. DOI: 10.1038/s41929-018-0045-1. 77, 97

Meichsner J, Schmidt M, Schneider R, Wagner H–E (2012) *Nonthermal Plasma Chemistry and Physics*. CRC Press. DOI: 10.1201/b12956. 2

Meinlschmidt P, Ueberham E, Lehmann J, Reineke K, Schlüter O, Schweiggert–Weisz U, Eisner P (2016) The effects of pulsed ultraviolet light, cold atmospheric pressure plasma, and gamma–irradiation on the immunoreactivity of soy protein isolate. *Innovative Food Science and Emerging Technologies* 38: 374–383. DOI: 10.1016/j.ifset.2016.06.007. 32

Meirow D, Biederman H, Anderson RA, Wallace WHB (2010) Toxicity of chemotherapy and radiation on female reproduction. *Clinical Obstetrics and Gynecology* 53: 727–739. DOI: 10.1097/GRF.0b013e3181f96b54. 57

Mendis D, Rosenberg M, Azam F (2000) A note on the possible electrostatic disruption of bacteria. *IEEE Transactions on Plasma Science* 28: 1304–1306. DOI: 10.1109/27.893321. 26, 37

Meng J, Nie W, Zhang K, Xu F, Ding X, Wang S, Qiu Y (2018) Enhancing electrochemical performance of graphene fiber–based supercapacitors by plasma treatment. *ACS Applied Materials and Interfaces* 10: 13652–13659. DOI: 10.1021/acsami.8b04438. 99, 100

Menon U, Poelman H, Bliznuk V, Galvita VV, Poelman D, Marin GB (2012) Nature of the active sites for the total oxidation of toluene by CuOCeO2/Al2O3. *Journal of Catalysis* 295: 91–103. DOI: 10.1016/j.jcat.2012.07.026. 88

Merche D, De Vos C, Baneton J, Dufour T, Godet S, Reniers F (2018) Synthesis and/or grafting of noble metal nanoparticles by microplasma and by atmospheric plasma torch. *arXiv preprint arXiv*:180202257. 80

Merche D, Vandencasteele N, Reniers F (2012) Atmospheric plasmas for thin film deposition: A critical review. *Thin Solid Films* 520: 4219–4236. DOI: 10.1016/j.tsf.2012.01.026. 104

Mertz G, Delmée M, Bardon J, Martin A, Ruch D, Fouquet T, Garreau S, Airoudj A, Marguier A, Ploux L (2018) Atmospheric pressure plasma co-polymerization of two acrylate precursors: Toward the control of wetting properties. *Plasma Processes and Polymers* 15: 1800073. DOI: 10.1002/ppap.201800073. 71

Metelmann H–R, Nedrelow DS, Seebauer C, Schuster M, von Woedtke T, Weltmann K–D, Kindler S, Metelmann PH, Finkelstein SE, Von Hoff DD (2015) Head and neck cancer treatment and physical plasma. *Clinical Plasma Medicine* 3: 17–23. DOI: 10.1016/j.cpme.2015.02.001. 53

Metelmann H–R, Seebauer C, Miller V, Fridman A, Bauer G, Graves DB, Pouvesle J–M, Rutkowski R, Schuster M, Bekeschus S (2018a) Clinical experience with cold plasma in the treatment of locally advanced head and neck cancer. *Clinical Plasma Medicine* 9: 6–13. DOI: 10.1016/j.cpme.2017.09.001. 53

Metelmann H–R, Von Woedtke T, Weltmann K–D (2018b) *Comprehensive Clinical Plasma Medicine: Cold Physical Plasma for Medical Application.* Springer. DOI: 10.1007/978-3-319-67627-2. 35

Miao W, Nyaisaba BM, Koddy JK, Chen M, Hatab S, Deng S (2020) Effect of cold atmospheric plasma on the physicochemical and functional properties of myofibrillar protein from Alaska pollock (Theragra chalcogramma). *International Journal of Food Science and Technology* 55: 517–525. DOI: 10.1111/ijfs.14295. 23

Minke C, Turek T (2018) Materials, system designs and modelling approaches in techno–economic assessment of all–vanadium redox flow batteries–A review. *Journal of Power Sources* 376: 66–81. DOI: 10.1016/j.jpowsour.2017.11.058. 98

Mirpour S, Piroozmand S, Soleimani N, Faharani NJ, Ghomi H, Eskandari HF, Sharifi AM, Mirpour S, Eftekhari M, Nikkhah M (2016) Utilizing the micron sized non–thermal atmospheric pressure plasma inside the animal body for the tumor treatment application. *Scientific Reports* 6: 29048. DOI: 10.1038/srep29048. 44

Misra N, Jo C (2017) Applications of cold plasma technology for microbiological safety in meat industry. *Trends in Food Science and Technology* 64: 74–86. DOI: 10.1016/j.tifs.2017.04.005. 37

Misra N, Koubaa M, Roohinejad S, Juliano P, Alpas H, Inacio RS, Saraiva JA, Barba FJ (2017) Landmarks in the historical development of twenty first century food processing technologies. *Food Research International* 97: 318–339. DOI: 10.1016/j.foodres.2017.05.001. 23, 30

Misra N, Pankaj S, Segat A, Ishikawa K (2016a) Cold plasma interactions with enzymes in foods and model systems. *Trends in Food Science and Technology* 55: 39–47. DOI: 10.1016/j.tifs.2016.07.001. 33, 37

Misra N, Schlüter O, Cullen PJ (2016b) *Cold Plasma in Food and Agriculture: Fundamentals and Applications*. Academic Press. DOI: 10.1016/B978-0-12-801365-6.00001-9. 23, 33, 103

Misra N, Yadav B, Roopesh M, Jo C (2019a) Cold plasma for effective fungal and mycotoxin control in foods: mechanisms, inactivation effects, and applications. *Comprehensive Reviews in Food Science and Food Safety* 18: 106–120. DOI: 10.1111/1541-4337.12398. 31

Misra N, Yepez X, Xu L, Keener K (2019b) In–package cold plasma technologies. *Journal of Food Engineering* 244: 21–31. DOI: 10.1016/j.jfoodeng.2018.09.019. 103

Mitra A, Li Y–F, Klämpfl TG, Shimizu T, Jeon J, Morfill GE, Zimmermann JL (2014) Inactivation of surface–borne microorganisms and increased germination of seed specimen by cold atmospheric plasma. *Food and Bioprocess Technology* 7: 645–653. DOI: 10.1007/s11947-013-1126-4. 24, 27

Mitra S, Nguyen LN, Akter M, Park G, Choi EH, Kaushik NK (2019) Impact of ROS generated by chemical, physical, and plasma techniques on cancer attenuation. *Cancers* 11: 1030. DOI: 10.3390/cancers11071030. 52

Mizuno A (2007) Industrial applications of atmospheric non–thermal plasma in environmental remediation. *Plasma Physics and Controlled Fusion* 49: A1. DOI: 10.1088/0741-3335/49/5A/S01. 90

Mo J, Zhang Y, Xu Q, Lamson JJ, Zhao R (2009a) Photocatalytic purification of volatile organic compounds in indoor air: a literature review. *Atmospheric Environment* 43: 2229–2246. DOI: 10.1016/j.atmosenv.2009.01.034. 90

Mo J, Zhang Y, Xu Q, Zhu Y, Lamson JJ, Zhao R (2009b) Determination and risk assessment of by–products resulting from photocatalytic oxidation of toluene. *Applied Catalysis B: Environmental* 89: 570–576. DOI: 10.1016/j.apcatb.2009.01.015. 90

Moghaddam S, Pengwang E, Jiang Y–B, Garcia AR, Burnett DJ, Brinker CJ, Masel RI, Shannon MA (2010) An inorganic–organic proton exchange membrane for fuel cells with a controlled nanoscale pore structure. *Nature Nanotechnology* 5: 230–236. DOI: 10.1038/nnano.2010.13. 79

Mohammadi S, Imani S, Dorranian D, Tirgari S, Shojaee M (2015) The effect of non–thermal plasma to control of stored product pests and changes in some characters of wheat materials. *Food Engineering Reviews* 7: 150–156. 31

Moisan M, Barbeau J, Moreau S, Pelletier J, Tabrizian M, L'H Y (2001) Low–temperature sterilization using gas plasmas: a review of the experiments and an analysis of the inactivation mechanisms. *International Journal of Pharmaceutics* 226: 1–21. DOI: 10.1016/S0378-5173(01)00752-9. 37

Molina R, López–Santos C, Gómez–Ramírez A, Vílchez A, Espinós JP, González–Elipe AR (2018) Influence of irrigation conditions in the germination of plasma treated Nasturtium seeds. *Scientific Reports* 8: 1–11. DOI: 10.1038/s41598-018-34801-0. 26, 27

Moljord K, Magnoux P, Guisnet M (1995) Coking, aging and regeneration of zeolites: XVI. influence of the composition of HY zeolites on the removal of coke through oxidative treatment. *Applied Catalysis A: General* 121: 245–259. DOI: 10.1016/0926-860X(94)00211-8. 76

Monthioux M, Allouche H, Jacobsen RL (2006) Chemical vapour deposition of pyrolytic carbon on carbon nanotubes: Part 3: Growth mechanisms. *Carbon* 44: 3183–3194. DOI: 10.1016/j.carbon.2006.07.001. 80

Moon A–Y, Noh S, Moon SY, You S (2016) Feasibility study of atmospheric–pressure plasma treated air gas package for grape's shelf–life improvement. *Current Applied Physics* 16: 440–445. DOI: 10.1016/j.cap.2016.01.007. 31

Moreau M, Orange N, Feuilloley M (2008) Non–thermal plasma technologies: new tools for bio–decontamination. *Biotechnology Advances* 26: 610–617. DOI: h10.1016/j.biotechadv.2008.08.001. 8, 37

Moreno–Couranjou M, Monthioux M, Gonzalez–Aguilar J, Fulcheri L (2009) A non–thermal plasma process for the gas phase synthesis of carbon nanoparticles. *Carbon* 47: 2310–2321. DOI: 10.1016/j.carbon.2009.04.003. 79, 80

Morozov AI (2012) *Introduction to Plasma Dynamics*. CRC Press. DOI: 10.1201/b13929. 1

Mošovská S, Medvecká V, Halászová N, Ďurina P, Valík Ľ, Mikulajová A, Zahoranová A (2018) Cold atmospheric pressure ambient air plasma inhibition of pathogenic bacteria on the surface of black pepper. *Food Research International* 106: 862–869. DOI: 10.1016/j.foodres.2018.01.066. 24

Mott–Smith HM (1971) History of "plasmas". *Nature* 233: 219–219. DOI: 10.1038/233219a0. 1

Moutiq R, Misra N, Mendonca A, Keener K (2020) In–package decontamination of chicken breast using cold plasma technology: Microbial, quality and storage studies. *Meat Science* 159: 107942. DOI: 10.1016/j.meatsci.2019.107942. 23

Murakami T, Niemi K, Gans T, O'Connell D, Graham WG (2012) Chemical kinetics and reactive species in atmospheric pressure helium–oxygen plasmas with humid–air impurities. *Plasma Sources Science and Technology* 22: 015003. DOI: 10.1088/0963-0252/22/1/015003. 15

Murawski K (2002) *Analytical and Numerical Methods for Wave Propagation in Fluid Media*. World Scientific. DOI: 10.1142/5092. 1

Muzumdar S, Hiebert H, Haertel E, Greenwald MB–Y, Bloch W, Werner S, Schäfer M (2019) Nrf2–Mediated Expansion of Pilosebaceous Cells Accelerates Cutaneous Wound Healing. *The American Journal of Pathology* 189: 568–579. DOI: 10.1016/j.ajpath.2018.11.017. 42

Naciri M, Dowling D, Al-Rubeai M (2014) Differential sensitivity of mammalian cell lines to non–thermal atmospheric plasma. *Plasma Processes and Polymers* 11: 391–400. DOI: 10.1002/ppap.201300118. 47

Nasonova A, Pham HC, Kim D–J, Kim K–S (2010) NO and SO2 removal in non–thermal plasma reactor packed with glass beads–TiO$_2$ thin film coated by PCVD process. *Chemical Engineering Journal* 156: 557–561. DOI: 10.1016/j.cej.2009.04.037. 89

Naujokat H, Harder S, Schulz LY, Wiltfang J, Flörke C, Açil Y (2019) Surface conditioning with cold argon plasma and its effect on the osseointegration of dental implants in miniature pigs. *Journal of Cranio–Maxillofacial Surgery* 47: 484–490. DOI: 10.1016/j.jcms.2018.12.011. 53

Ndjeri M, Pensel A, Peulon S, Haldys V, Desmazières B, Chaussé A (2013) Degradation of glyphosate and AMPA (amino methylphosphonic acid) solutions by thin films of birnessite electrodeposited: A new design of material for remediation processes? *Colloids and Surfaces A: Physicochemical and Engineering Aspects* 435: 154–169. DOI: 10.1016/j. colsurfa.2013.01.022. 68

Nejat F, Nabavi N–S, Nejat M–A, Aghamollaei H, Jadidi K (2019) Safety evaluation of the plasma on ocular surface tissue: An animal study and histopathological findings. *Clinical Plasma Medicine* 14: 100084. DOI: 10.1016/j.cpme.2019.100084. 53

Neyens E, Baeyens J (2003) A review of classic Fenton's peroxidation as an advanced oxidation technique. *Journal of Hazardous Materials* 98: 33–50. DOI: 10.1016/S0304-3894(02)00282-0. 20

Neyts EC (2016) Plasma–surface interactions in plasma catalysis. *Plasma Chemistry and Plasma Processing* 36: 185–212. DOI: 10.1007/s11090-015-9662-5. 92

Neyts EC, Ostrikov K, Sunkara MK, Bogaerts A (2015) Plasma catalysis: synergistic effects at the nanoscale. *Chemical Reviews* 115: 13408–13446. DOI: 10.1021/acs.chemrev.5b00362. 5, 6, 77, 83, 87, 93, 103, 105

Ngo Thi MH, Shao PL, Liao JD, Lin CCK, Yip HK (2014) Enhancement of angiogenesis and epithelialization processes in mice with burn wounds through ROS/RNS signals generated by non-thermal N_2/Ar micro-plasma. *Plasma Processes and Polymers* 11: 1076–1088. DOI: 10.1002/ppap.201400072. 41

Nguyen DV, Ho PQ, Pham TV, Nguyen TV, Kim L, Van Nguyen D, Ho PQ, Van Pham T, Van Nguyen T, Kim L (2018) Treatment of surface water using cold plasma for domestic water supply. *Environmental Engineering Research* 24: 412–417. DOI: /10.4491/eer.2018.215. 7

Nguyen L, Hang M, Wang W, Tian Y, Wang L, McCarthy TJ, Chen W (2014) Simple and improved approaches to long–lasting, hydrophilic silicones derived from commercially available precursors. *ACS Applied Materials and Interfaces* 6: 22876–22883. DOI: 10.1021/am507152d. 74

Ni C, Carolan D, Rocks C, Hui J, Fang Z, Padmanaban DB, Ni J, Xie D, Maguire P, Irvine JT (2018) Microplasma–assisted electrochemical synthesis of Co_3O_4 nanoparticles in absolute ethanol for energy applications. *Green Chemistry* 20: 2101–2109. DOI: 10.1039/C8GC00200B. 80

Niemira BA (2012) Cold plasma decontamination of foods. *Annual Review of Food Science and Technology* 3: 125–142. DOI: 10.1146/annurev-food-022811-101132. 30, 103

Niemira BA (2014) Decontamination of foods by cold plasma. In: *Emerging Technologies for Food Processing*, pp. 327–333. Elsevier. DOI: 10.1016/B978-0-12-411479-1.00018-8. 30

Nikiforov AY, Leys C, Gonzalez M, Walsh J (2015) Electron density measurement in atmospheric pressure plasma jets: Stark broadening of hydrogenated and non–hydrogenated lines.

Plasma Sources Science and Technology 24: 034001. DOI: 10.1088/0963-0252/24/3/034001. 12

Nimlos MR, Wolfrum EJ, Brewer ML, Fennell JA, Bintner G (1996) Gas–phase heterogeneous photocatalytic oxidation of ethanol: pathways and kinetic modeling. *Environmental Science and Technology* 30: 3102–3110. DOI: 10.1021/es9602298. 92

Nishime T, Borges A, Koga–Ito C, Machida M, Hein L, Kostov K (2017) Non–thermal atmospheric pressure plasma jet applied to inactivation of different microorganisms. *Surface and Coatings Technology* 312: 19–24. DOI: 10.1016/j.surfcoat.2016.07.076. 8

Nizio M, Albarazi A, Cavadias S, Amouroux J, Galvez ME, Da Costa P (2016) Hybrid plasma–catalytic methanation of CO_2 at low temperature over ceria zirconia supported Ni catalysts. *International Journal of Hydrogen Energy* 41: 11584–11592. DOI: /10.1016/j.ijhydene.2016.02.020. 87

Nolan H, Sun D, Falzon BG, Chakrabarti S, Padmanaba DB, Maguire P, Mariotti D, Yu T, Jones D, Andrews G (2018) Metal nanoparticle-hydrogel nanocomposites for biomedical applications–An atmospheric pressure plasma synthesis approach. *Plasma Processes and Polymers* 15: 1800112. DOI: 10.1002/ppap.201800112. 77, 78, 80

Nold BT, Zenker M, Selig P (2018) Test electrode for argon plasma coagulation medical instrument. Google Patents. 7

Norsic C, Tatibouët J–M, Batiot–Dupeyrat C, Fourré E (2018) Methanol oxidation in dry and humid air by dielectric barrier discharge plasma combined with MnO_2–CuO based catalysts. *Chemical Engineering Journal* 347: 944–952. DOI: 10.1016/j.cej.2018.04.065. 92

Novák I, Popelka A, Luyt A, Chehimi M, Špírková M, Janigová I, Kleinová A, Stopka P, Šlouf M, Vanko V (2013) Adhesive properties of polyester treated by cold plasma in oxygen and nitrogen atmospheres. *Surface and Coatings Technology* 235: 407–416. DOI: 10.1016/j.surfcoat.2013.07.057. 13

Novák I, Valentin M, Špitalský Z, Popelka A, Sestak J, Krupa I (2018) Superhydrophobic polyester/cotton fabrics modified by barrier discharge plasma and organosilanes. *Polymer–Plastics Technology and Engineering* 57: 440–448. DOI: 10.1080/03602559.2017.1289397. 74

Nozaki T, Okazaki K (2013) Non–thermal plasma catalysis of methane: Principles, energy efficiency, and applications. *Catalysis Today* 211: 29–38. DOI: 10.1016/j.cattod.2013.04.002. 87, 93

Nozaki T, Tsukijihara H, Fukui W, Okazaki K (2007) Kinetic analysis of the catalyst and non–thermal plasma hybrid reaction for methane steam reforming. *Energy and Fuels* 21: 2525–2530. DOI: 10.1021/ef070117+. 88

Obradović BM, Sretenović GB, Kuraica MM (2011) A dual–use of DBD plasma for simultaneous NO_x and SO_2 removal from coal–combustion flue gas. *Journal of Hazardous Materials* 185: 1280–1286. DOI: 10.1016/j.jhazmat.2010.10.043. 62

Oda T (2003) Non–thermal plasma processing for environmental protection: decomposition of dilute VOCs in air. *Journal of Electrostatics* 57: 293–311. DOI: 10.1016/S0304-3886(02)00179-1. 92

Oehmigen K, Hähnel M, Brandenburg R, Wilke C, Weltmann KD, Von Woedtke T (2010) The role of acidification for antimicrobial activity of atmospheric pressure plasma in liquids. *Plasma Processes and Polymers* 7: 250–257. DOI: 10.1002/ppap.200900077. 28

Oehrlein GS, Hamaguchi S (2018) Foundations of low–temperature plasma enhanced materials synthesis and etching. *Plasma Sources Science and Technology* 27: 023001. DOI: 10.1088/1361-6595/aaa86c. 71

Okazaki Y, Wang Y, Tanaka H, Mizuno M, Nakamura K, Kajiyama H, Kano H, Uchida K, Kikkawa F, Hori M (2014) Direct exposure of non–equilibrium atmospheric pressure plasma confers simultaneous oxidative and ultraviolet modifications in biomolecules. *Journal of Clinical Biochemistry and Nutrition* 55: 207–215. DOI: 10.3164/jcbn.14-40. 31

Olaimat AN, Holley RA (2012) Factors influencing the microbial safety of fresh produce: a review. *Food Microbiology* 32: 1–19. DOI: 10.1016/j.fm.2012.04.016. 30

Ölmez H, Temur S (2010) Effects of different sanitizing treatments on biofilms and attachment of Escherichia coli and Listeria monocytogenes on green leaf lettuce. *LWT-Food Science and Technology* 43: 964–970. DOI: 10.1016/j.lwt.2010.02.005. 30

Ono R, Tokumitsu Y (2015) Selective production of atomic oxygen by laser photolysis as a tool for studying the effect of atomic oxygen in *Plasma Medicine. Journal of Physics D: Applied Physics* 48: 275201. DOI: 10.1088/0022-3727/48/27/275201. 5

Oonnittan A, Shrestha RA, Sillanpää M (2009) Removal of hexachlorobenzene from soil by electrokinetically enhanced chemical oxidation. *Journal of Hazardous Materials* 162: 989–993. DOI: 10.1016/j.jhazmat.2008.05.132. 66

Organization WH (2018) WHO expert consultation on rabies: third report. World Health Organization. 60

Ostrikov K (2005) Colloquium: Reactive plasmas as a versatile nanofabrication tool. *Reviews of Modern Physics* 77: 489. DOI: 10.1103/RevModPhys.77.489. 2

Ostrikov K, Neyts E, Meyyappan M (2013) Plasma nanoscience: from nano–solids in plasmas to nano–plasmas in solids. *Advances in Physics* 62: 113–224. DOI: 10.1080/00018732.2013.808047. 7

Ouyang B, Zhang Y, Zhang Z, Fan HJ, Rawat RS (2017) Nitrogen-plasma-activated hierarchical nickel nitride nanocorals for energy applications. *Small* 13: 1604265. DOI: 10.1002/smll.201604265. 71

Ozkan A, Dufour T, Arnoult G, De Keyzer P, Bogaerts A, Reniers F (2015) CO2–CH4 conversion and syngas formation at atmospheric pressure using a multi–electrode dielectric barrier discharge. *Journal of CO₂ Utilization* 9: 74–81. DOI: 10.1016/j.jcou.2015.01.002. 96

Pacher P, Beckman JS, Liaudet L (2007) Nitric oxide and peroxynitrite in health and disease. *Physiological Reviews* 87: 315–424. DOI: 10.1152/physrev.00029.2006. 49

Padan E, Schuldiner S (1987) Intracellular pH and membrane potential as regulators in the prokaryotic cell. *The Journal of Membrane Biology* 95: 189–198. DOI: 10.1007/BF01869481. 37

Pandiyaraj K, Ramkumar M, Kumar AA, Padmanabhan P, Pichumani M, Bendavid A, Cools P, De Geyter N, Morent R, Kumar V (2019) Evaluation of surface properties of low density polyethylene (LDPE) films tailored by atmospheric pressure non–thermal plasma (APNTP) assisted co–polymerization and immobilization of chitosan for improvement of antifouling properties. *Materials Science and Engineering: C* 94: 150–160. DOI: 10.1016/j.msec.2018.08.062. 71

Pangilinan CDC, Kurniawan W, Hinode H (2016) Effect of MnOx/TiO2 Oxidation state on ozone concentration in a nonthermal plasma–driven catalysis reactor. *Ozone: Science and Engineering* 38: 156–162. DOI: 10.1080/01919512.2015.1091720. 92

Pankaj S, Misra N, Cullen P (2013) Kinetics of tomato peroxidase inactivation by atmospheric pressure cold plasma based on dielectric barrier discharge. *Innovative Food Science and Emerging Technologies* 19: 153–157. DOI: 10.1016/j.ifset.2013.03.001. 30

Pappas D (2011) Status and potential of atmospheric plasma processing of materials. *Journal of Vacuum Science and Technology A: Vacuum, Surfaces, and Films* 29: 020801. DOI: 10.1116/1.3559547. 3

Parameswaran N, Patial S (2010) Tumor necrosis factor–α signaling in macrophages. *Critical Reviews™ in Eukaryotic Gene Expression* 20. DOI: 10.1615/CritRevEukarGeneExpr.v20.i2.10. 42

Park BJ, Lee D, Park J–C, Lee I–S, Lee K–Y, Hyun S, Chun M–S, Chung K–H (2003) Sterilization using a microwave–induced argon plasma system at atmospheric pressure. *Physics of Plasmas* 10: 4539–4544. DOI: 10.1063/1.1613655. 73

Park C–S, Jung EY, Jang HJ, Bae GT, Shin BJ, Tae H–S (2019) Synthesis and properties of plasma–polymerized methyl methacrylate via the atmospheric pressure plasma polymerization technique. *Polymers* 11: 396. DOI: 10.3390/polym11030396. 71

Park CH, Lee JS, Kim JH, Kim D–K, Lee OJ, Ju HW, Moon BM, Cho JH, Kim MH, Sun PP (2014) Wound healing with nonthermal microplasma jets generated in arrays of hourglass microcavity devices. *Journal of Physics D: Applied Physics* 47: 435402. DOI: 10.1088/0022-3727/47/43/435402. 4

Park CH, Lee SY, Hwang DS, Shin DW, Cho DH, Lee KH, Kim T–W, Kim T–W, Lee M, Kim D–S (2016) Nanocrack–regulated self–humidifying membranes. *Nature* 532: 480–483. DOI: 10.1038/nature17634. 71, 77, 78

Park CW, Byeon JH, Yoon KY, Park JH, Hwang J (2011) Simultaneous removal of odors, airborne particles, and bioaerosols in a municipal composting facility by dielectric barrier discharge. *Separation and Purification Technology* 77: 87–93. DOI: 10.1016/j.seppur.2010.11.024. 62

Park DP, Davis K, Gilani S, Alonzo C–A, Dobrynin D, Friedman G, Fridman A, Rabinovich A, Fridman G (2013) Reactive nitrogen species produced in water by non–equilibrium plasma increase plant growth rate and nutritional yield. *Current Applied Physics* 13: S19–S29. DOI: 10.1016/j.cap.2012.12.019. 31

Park G, Park S, Choi M, Koo I, Byun J, Hong J, Sim J, Collins G, Lee J (2012) Atmospheric–pressure plasma sources for biomedical applications. *Plasma Sources Science and Technology* 21: 043001. DOI: 10.1088/0963-0252/21/4/043001. 9, 35

Park J, Huh J–E, Lee S–E, Lee J, Lee WH, Lim K–H, Kim YS (2018) Effective atmospheric–pressure plasma treatment toward high–performance solution–processed oxide thin–film transistors. *ACS Applied Materials and Interfaces* 10: 30581–30586. DOI: 10.1021/acsami.8b11111. 71

Park JH, Kumar N, Uhm HS, Lee W, Choi EH, Attri P (2015) Effect of nanosecond–pulsed plasma on the structural modification of biomolecules. *RSC Advances* 5: 47300–47308. DOI: 10.1039/C5RA04993H. 31

Park JY, Lee YN (1988) Solubility and decomposition kinetics of nitrous acid in aqueous solution. *The Journal of Physical Chemistry* 92: 6294–6302. DOI: 10.1021/j100333a025. 18, 19

Park S–J, Choi J, Park GY, Lee S–K, Cho Y, Yun JI, Jeon S, Kim KT, Lee JK, Sim J–Y (2010) Inactivation of S. mutans using an atmospheric plasma driven by a palm–size–integrated microwave power module. *IEEE Transactions on Plasma Science* 38: 1956–1962. DOI: 10.1109/TPS.2010.2048112. 8

Partecke LI, Evert K, Haugk J, Doering F, Normann L, Diedrich S, Weiss F–U, Evert M, Huebner NO, Guenther C (2012) Tissue tolerable plasma (TTP) induces apoptosis in pancreatic cancer cells *in vitro* and *in vivo*. *BMC Cancer* 12: 1–10. DOI: 10.1186/1471-2407-12-473. 5

Pastor–Pérez L, Belda–Alcázar V, Marini C, Pastor–Blas MM, Sepúlveda–Escribano A, Ramos–Fernandez EV (2018) Effect of cold Ar plasma treatment on the catalytic performance of Pt/CeO_2 in water–gas shift reaction (WGS). *Applied Catalysis B: Environmental* 225: 121–127. DOI: 10.1016/j.apcatb.2017.11.065. 71

Patange A, Boehm D, Bueno–Ferrer C, Cullen P, Bourke P (2017) Controlling Brochothrix thermosphacta as a spoilage risk using in–package atmospheric cold plasma. *Food Microbiology* 66: 48–54. DOI: 10.1016/j.fm.2017.04.002. 31

Patange A, Boehm D, Giltrap M, Lu P, Cullen P, Bourke P (2018) Assessment of the disinfection capacity and eco–toxicol 90 ogical impact of atmospheric cold plasma for treatment of food industry effluents. *Science of the Total Environment* 631: 298–307. DOI: 10.1016/j.scitotenv.2018.02.269. 63

Patil B, Cherkasov N, Lang J, Ibhadon A, Hessel V, Wang Q (2016) Low temperature plasma–catalytic NO_x synthesis in a packed DBD reactor: Effect of support materials and supported active metal oxides. *Applied Catalysis B: Environmental* 194: 123–133. DOI: 10.1016/j.apcatb.2016.04.055. 90

Paulmier T, Fulcheri L (2005) Use of non–thermal plasma for hydrocarbon reforming. *Chemical Engineering Journal* 106: 59–71. DOI: 10.1016/j.cej.2004.09.005. 84

Pavliňák D, Galmiz O, Zemánek M, Černák M (2018) Design and evaluation of plasma polymer deposition on hollow objects by electrical plasma generated from the liquid surface. *Plasma Processes and Polymers* 15: 1700183. DOI: 10.1002/ppap.201700183. 71

Peng M, Li L, Xiong J, Hua K, Wang S, Shao T (2017) Study on surface properties of polyamide 66 using atmospheric glow–like discharge plasma treatment. *Coatings* 7: 123. DOI: 10.3390/coatings7080123. 71

Peng P, Chen P, Addy M, Cheng Y, Anderson E, Zhou N, Schiappacasse C, Zhang Y, Chen D, Hatzenbeller R (2018a) Atmospheric plasma–assisted ammonia synthesis enhanced via synergistic catalytic absorption. *ACS Sustainable Chemistry and Engineering* 7: 100–104. DOI: 10.1021/acssuschemeng.8b03887. 77

Peng P, Chen P, Schiappacasse C, Zhou N, Anderson E, Chen D, Liu J, Cheng Y, Hatzenbeller R, Addy M (2018b) A review on the non–thermal plasma–assisted ammonia synthesis technologies. *Journal of Cleaner Production* 177: 597–609. DOI: 10.1016/j.jclepro.2017.12.229. 97

Peng P, Li Y, Cheng Y, Deng S, Chen P, Ruan R (2016) Atmospheric pressure ammonia synthesis using non–thermal plasma assisted catalysis. *Plasma Chemistry and Plasma Processing* 36: 1201–1210. DOI: 10.1007/s11090-016-9713-6. 93

Peng Y, Kajiyama H, Nakamura K, Utsumi F, Yoshikawa N, Tanaka H, Mizuno M, Toyokuni S, Hori M, Kikkawa F (2018c) Plasma–activated medium inhibits metastatic activities of ovarian cancer cells *in vitro* via repressing mapk pathway. *Clinical Plasma Medicine* 9: 41–42. DOI: 10.1016/j.cpme.2017.12.065. 47

Perathoner S, Gross S, Hensen EJ, Wessel H, Chraye H, Centi G (2017) Looking at the future of chemical production through the european roadmap on science and technology of catalysis the EU effort for a long-term vision. *ChemCatChem* 9: 904–909. DOI: 10.1002/cctc.201601641. 83

Pérez–Andrés JM, de Alba M, Harrison SM, Brunton NP, Cullen P, Tiwari BK (2020) Effects of cold atmospheric plasma on mackerel lipid and protein oxidation during storage. *LWT* 118: 108697. DOI: 10.1016/j.lwt.2019.108697. 31

Pérez–Gómez E, Jerkic M, Prieto M, Del Castillo G, Martin–Villar E, Letarte M, Bernabeu C, Pérez–Barriocanal F, Quintanilla M, López–Novoa JM (2014) Impaired wound repair in adult endoglin heterozygous mice associated with lower NO bioavailability. *Journal of Investigative Dermatology* 134: 247–255. DOI: 10.1038/jid.2013.263. 41

Perna A, Minutillo M, Jannelli E (2016) Hydrogen from intermittent renewable energy sources as gasification medium in integrated waste gasification combined cycle power plants: A performance comparison. *Energy* 94: 457–465. DOI: 10.1016/j.energy.2015.10.143. 96

Perni S, Liu DW, Shama G, Kong MG (2008) Cold atmospheric plasma decontamination of the pericarps of fruit. *Journal of Food Protection* 71: 302–308. DOI: 10.4315/0362-028X-71.2.302. 30

Person JC, Ham DO (1988) Removal of SO_2 and NO_x from stack gases by electron beam irradiation. *International Journal of Radiation Applications and Instrumentation Part C Radiation Physics and Chemistry* 31: 1–8. DOI: 10.1016/1359-0197(88)90103-8. 16

Petitpas G, Rollier J–D, Darmon A, Gonzalez–Aguilar J, Metkemeijer R, Fulcheri L (2007) A comparative study of non–thermal plasma assisted reforming technologies. *International Journal of Hydrogen Energy* 32: 2848–2867. DOI: 10.1016/j.ijhydene.2007.03.026. 84

Petrosillo N, Viceconte G, Ergonul O, Ippolito G, Petersen E (2020) COVID–19, SARS and MERS: are they closely related? *Clinical Microbiology and Infection.* DOI: 10.1016/j.cmi.2020.03.026. 38

Peyrous R, Pignolet P, Held B (1989) Kinetic simulation of gaseous species created by an electrical discharge in dry or humid oxygen. *Journal of Physics D: Applied Physics* 22: 1658. DOI: 10.1088/0022-3727/22/11/015. 15, 17

Phan KTK, Phan HT, Brennan CS, Phimolsiripol Y (2017a) Nonthermal plasma for pesticide and microbial elimination on fruits and vegetables: an overview. *International Journal of Food Science* and Technology 52: 2127–2137. DOI: 10.1111/ijfs.13509. 8

Phan LT, Yoon SM, Moon M–W (2017b) Plasma–based nanostructuring of polymers: A review. *Polymers* 9: 417. DOI: 10.3390/polym9090417. 80

Phelps A (1985) Tabulations of collision cross sections and calculated transport and reaction coefficients for electron collisions with O_2. *JILA Information Center Report* 28. 14, 15

Pignata C, D'angelo D, Fea E, Gilli G (2017) A review on microbiological decontamination of fresh produce with nonthermal plasma. *Journal of Applied Microbiology* 122: 1438–1455. DOI: 10.1111/jam.13412. 31

Pina–Perez M, Martinet D, Palacios–Gorba C, Ellert C, Beyrer M (2020) Low–energy short–term cold atmospheric plasma: Controlling the inactivation efficacy of bacterial spores in powders. *Food Research International* 130: 108921. DOI: 10.1016/j.foodres.2019.108921. 23

Pinela J, Ferreira IC (2017) Nonthermal physical technologies to decontaminate and extend the shelf–life of fruits and vegetables: Trends aiming at quality and safety. *Critical Reviews in Food Science and Nutrition* 57: 2095–2111. DOI: 10.1080/10408398.2015.1046547. 30

Pinto IS, Pacheco PH, Coelho JV, Lorençon E, Ardisson JD, Fabris JD, de Souza PP, Krambrock KW, Oliveira LC, Pereira MC (2012) Nanostructured δ–FeOOH: an efficient Fenton–like catalyst for the oxidation of organics in water. *Applied Catalysis B: Environmental* 119: 175–182. DOI: 10.1016/j.apcatb.2012.02.026. 86

Pipa A, Ionikh YZ, Chekishev V, Dünnbier M, Reuter S (2015) Resonance broadening of argon lines in a micro–scaled atmospheric pressure plasma jet (argon μAPPJ). *Applied Physics Letters* 106: 244104. DOI: 10.1063/1.4922730. 12

Popov N (2011) Fast gas heating in a nitrogen–oxygen discharge plasma: I. Kinetic mechanism. *Journal of Physics D: Applied Physics* 44: 285201. DOI: 10.1088/0022-3727/44/28/285201.15

Praveen K, Pious C, Thomas S, Grohens Y (2019) Relevance of plasma processing on polymeric materials and interfaces. In: *Non–Thermal Plasma Technology for Polymeric Materials*, pp. 1–21. Elsevier: 10.1016/B978-0-12-813152-7.00001-9. 71

Privat–Maldonado A, Bengtson C, Razzokov J, Smits E, Bogaerts A (2019a) Modifying the tumour microenvironment: Challenges and future perspectives for anticancer plasma treatments. *Cancers* 11: 1920. DOI: 10.3390/cancers11121920. 53

Privat–Maldonado A, Schmidt A, Lin A, Weltmann K–D, Wende K, Bogaerts A, Bekeschus S (2019b) ROS from physical plasmas: Redox chemistry for biomedical therapy. *Oxidative Medicine and Cellular Longevity* 2019. DOI: 10.1155/2019/9062098. 5, 35

Profili J, Rousselot S, Tomassi E, Briqueleur E, Aymé–Perrot D, Stafford L, Dollé M (2020) Toward more sustainable rechargeable aqueous batteries using plasma–treated cellulose–based li–ion electrodes. *ACS Sustainable Chemistry and Engineering* 8: 4728–4733. DOI: 10.1021/acssuschemeng.9b07125. 7, 101

Pryor WA, Squadrito GL (1995) The chemistry of peroxynitrite: a product from the reaction of nitric oxide with superoxide. *American Journal of Physiology–Lung Cellular and Molecular Physiology* 268: L699–L722. DOI: 10.1152/ajplung.1995.268.5.L699. 19

Puač N, Gherardi M, Shiratani M (2018) Plasma agriculture: A rapidly emerging field. *Plasma Processes and Polymers* 15: 1700174. DOI: 10.1002/ppap.201700174. 23, 28

Puligundla P, Kim J–W, Mok C (2017) Effect of corona discharge plasma jet treatment on de-contamination and sprouting of rapeseed (Brassica napus L.) seeds. *Food Control* 71: 376–382. DOI: 10.1016/j.foodcont.2016.07.021. 24 93

Puliyalil H, Cvelbar U (2016) Selective plasma etching of polymeric substrates for advanced applications. *Nanomaterials* 6: 108. DOI: 10.3390/nano6060108. 73

Puliyalil H, Jurković DL, Dasireddy VD, Likozar B (2018) A review of plasma–assisted catalytic conversion of gaseous carbon dioxide and methane into value–added platform chemicals and fuels. *RSC Advances* 8: 27481–27508. DOI: 10.1039/C8RA03146K. 93

Qin S, Wang M, Wang C, Jin Y, Yuan N, Wu Z, Zhang J (2018) Binder-free nanoparticulate coating of a polyethylene separator via a reactive atmospheric pressure plasma for lithium-ion batteries with improved performances. *Advanced Materials Interfaces* 5: 1800579. DOI: 10.1002/admi.201800579. 98, 100, 107

Qu G–Z, Li J, Liang D–L, Huang D–L, Qu D, Huang Y–M (2013) Surface modification of a granular activated carbon by dielectric barrier discharge plasma and its effects on pentachlorophenol adsorption. *Journal of Electrostatics* 71: 689–694. DOI: 10.1016/j.elstat.2013.03.013. 67

Rahemi N, Haghighi M, Babaluo AA, Jafari MF, Estifaee P (2013) Synthesis and physicochemical characterizations of Ni/Al_2O_3–ZrO_2 nanocatalyst prepared via impregnation method and treated with non–thermal plasma for CO_2 reforming of CH_4. *Journal of Industrial and Engineering Chemistry* 19: 1566–1576. DOI: 10.1016/j.jiec.2013.01.024. 84

Rahimnejad M, Adhami A, Darvari S, Zirepour A, Oh S–E (2015) Microbial fuel cell as new technology for bioelectricity generation: A review. *Alexandria Engineering Journal* 54: 745–756. DOI: 10.1016/j.aej.2015.03.031. 98

Raiser J, Zenker M (2006) Argon plasma coagulation for open surgical and endoscopic applications: state of the art. *Journal of Physics D: Applied Physics* 39: 3520. DOI: 10.1088/0022-3727/39/16/S10. 35

Raizer YP, Mokrov M (2013) Physical mechanisms of self–organization and formation of current patterns in gas discharges of the Townsend and glow types. *Physics of Plasmas* 20: 101604. DOI: 10.1063/1.4823460. 27

Raj L, Ide T, Gurkar AU, Foley M, Schenone M, Li X, Tolliday NJ, Golub TR, Carr SA, Shamji AF (2011) Selective killing of cancer cells by a small molecule targeting the stress response to ROS. *Nature* 475: 231–234. DOI: 10.1038/nature10167. 44

Rana S, Mehta D, Bansal V, Shivhare U, Yadav SK (2020) Atmospheric cold plasma (ACP) treatment improved in–package shelf–life of strawberry fruit. *Journal of Food Science and Technology* 57: 102–112. DOI: 10.1007/s13197-019-04035-7. 30

Rat V, Coudert J–F (2006) A simplified analytical model for dc plasma spray torch: influence of gas properties and experimental conditions. *Journal of Physics D: Applied Physics* 39: 4799. DOI: 10.1088/0022-3727/39/22/010. 13

Reddy PMK, Subrahmanyam C (2012) Green approach for wastewater treatment. Degradation and mineralization of aqueous organic pollutants by discharge plasma. *Industrial and Engineering Chemistry Research* 51: 11097–11103. DOI: 10.1021/ie301122p. 63

Reuter S, Winter J, Schmidt–Bleker A, Schroeder D, Lange H, Knake N, Schulz–Von Der Gathen V, Weltmann K (2012) Atomic oxygen in a cold argon plasma jet: TALIF spectroscopy in ambient air with modelling and measurements of ambient species diffusion. *Plasma Sources Science and Technology* 21: 024005. DOI: 10.1088/0963-0252/21/2/024005. 13

Ribas A, Wolchok JD (2018) Cancer immunotherapy using checkpoint blockade. *Science* 359: 1350–1355. DOI: 10.1126/science.aar4060. 52

Ricard A, Décomps P, Massines F (1999) Kinetics of radiative species in helium pulsed discharge at atmospheric pressure. *Surface and Coatings Technology* 112: 1–4. DOI: 10.1016/S0257-8972(98)00797-X. 14, 17, 18

Ricciardolo FL, Sterk PJ, Gaston B, Folkerts G (2004) Nitric oxide in health and disease of the respiratory system. *Physiological Reviews* 84: 731–765. DOI: 10.1152/physrev.00034.2003. 55

Richard JL (2007) Some major mycotoxins and their mycotoxicoses—An overview. *International Journal of Food Microbiology* 119: 3–10. DOI: 10.1016/j.ijfoodmicro.2007.07.019. 23

Riedl SJ, Shi Y (2004) Molecular mechanisms of caspase regulation during apoptosis. *Na* 95 *ture Reviews Molecular Cell Biology* 5: 897–907. DOI: 10.1038/nrm1496. 49

Riès D, Dilecce G, Robert E, Ambrico P, Dozias S, Pouvesle JM (2014) LIF and fast imaging plasma jet characterization relevant for NTP biomedical applications. *Journal of Physics D: Applied Physics* 47: 275401. DOI: 10.1088/0022-3727/47/27/275401. 10

Rodríguez–Monroy C, Mármol–Acitores G, Nilsson–Cifuentes G (2018) Electricity generation in Chile using non–conventional renewable energy sources–A focus on biomass. *Renewable and Sustainable Energy Reviews* 81: 937–945. DOI: 10.1016/j.rser.2017.08.059. 95

Roland U, Holzer F, Kopinke F–D (2005) Combination of non–thermal plasma and heterogeneous catalysis for oxidation of volatile organic compounds: Part 2. Ozone decomposition and deactivation of γ–Al_2O_3. *Applied Catalysis B: Environmental* 58: 217–226. DOI: 10.1016/j.apcatb.2004.11.024. 83

Roth C, Gerullis S, Beier O, Pfuch A, Hartmann A, Grünler B (2018) Atmospheric pressure plasma as a tool for decontamination and functionalization of wooden surfaces. *MS&E* 364: 012077. DOI: 10.1088/1757-899X/364/1/012077. 71, 75

Rowe B, Vallee F, Queffelec J, Gomet J, Morlais M (1988) The yield of oxygen and hydrogen atoms through dissociative recombination of H_2O^+ ions with electrons. *The Journal of Chemical Physics* 88: 845–850. DOI: 10.1063/1.454164. 18

Ru X, Yao X (2014) TRPM2: a multifunctional ion channel for oxidative stress sensing. *Acta Physiologica Sinica* 66: 7. 50

Rykowska I, Wasiak W (2015) Research trends on emerging environment pollutants–a review. *Open Chemistry* 1. DOI: 10.1515/chem-2015-0151. 59, 66

Sachs M, Schmidt J, Peukert W, Wirth K–E (2018) Treatment of polymer powders by combining an atmospheric plasma jet and a fluidized bed reactor. *Powder Technology* 325: 490–497. DOI: 10.1016/j.powtec.2017.11.016. 74

Sadhu S, Thirumdas R, Deshmukh R, Annapure U (2017) Influence of cold plasma on the enzymatic activity in germinating mung beans (Vigna radiate). *LWT* 78: 97–104. DOI: 10.1016/j.lwt.2016.12.026. 24

Sahoo G, Polaki S, Ghosh S, Krishna N, Kamruddin M (2018) Temporal–stability of plasma functionalized vertical graphene electrodes for charge storage. *Journal of Power Sources* 401: 37–48. DOI: 10.1016/j.jpowsour.2018.08.071. 99

Sakiyama Y, Graves DB, Chang H–W, Shimizu T, Morfill GE (2012) Plasma chemistry model of surface microdischarge in humid air and dynamics of reactive neutral species. *Journal of Physics D: Applied Physics* 45: 425201. DOI: 10.1088/0022-3727/45/42/425201. 15, 16, 18

Sakudo A, Misawa T, Yagyu Y (2020) Equipment design for cold plasma disinfection of food products. In: *Advances in Cold Plasma Applications for Food Safety and Preservation,* pp. 289–307. Elsevier. DOI: 10.1016/B978-0-12-814921-8.00010-4. 7

Salanne M, Rotenberg B, Naoi K, Kaneko K, Taberna P–L, Grey CP, Dunn B, Simon P (2016) Efficient storage mechanisms for building better supercapacitors. *Nature Energy* 1: 1–10. DOI: 10.1038/nenergy.2016.70. 99

Salem W, Leitner DR, Zingl FG, Schratter G, Prassl R, Goessler W, Reidl J, Schild S (2015) Antibacterial activity of silver and zinc nanoparticles against Vibrio cholerae and enterotoxic Escherichia coli. *International Journal of Medical Microbiology* 305: 85–95. DOI: 10.1016/j.ijmm.2014.11.005. 77

Samukawa S, Hori M, Rauf S, Tachibana K, Bruggeman P, Kroesen G, Whitehead JC, Murphy AB, Gutsol AF, Starikovskaia S (2012) The 2012 plasma roadmap. *Journal of Physics D: Applied Physics* 45: 253001. DOI: 10.1088/0022-3727/45/25/253001. 6, 103

Sanbhal N, Mao Y, Sun G, Xu RF, Zhang Q, Wang L (2018) Surface modification of polypropylene mesh devices with cyclodextrin via cold plasma for hernia repair: Characterization and antibacterial properties. *Applied Surface Science* 439: 749–759. DOI: 10.1016/j.apsusc.2017.12.192. 7

Sandu C, Popescu M, Rosales E, Pazos M, Lazar G, Sanromán MÁ (2017) Electrokinetic oxidant soil flushing: a solution for in situ remediation of hydrocarbons polluted soils. *Journal of Electroanalytical Chemistry* 799: 1–8. DOI: 10.1016/j.jelechem.2017.05.036. 66

Sarinont T, Amano T, Attri P, Koga K, Hayashi N, Shiratani M (2016) Effects of plasma irradiation using various feeding gases on growth of Raphanus sativus L. *Archives of Biochemistry and Biophysics* 605: 129–140. DOI: 10.1016/j.abb.2016.03.024. 28, 29

Scally L, Lalor J, Cullen PJ, Milosavljević V (2017) Impact of atmospheric pressure nonequilibrium plasma discharge on polymer surface metrology. *Journal of Vacuum Science and Technology A: Vacuum, Surfaces, and Films* 35: 03E105. DOI: 10.1116/1.4978254. 74

Schlegel J, Köritzer J, Boxhammer V (2013) Plasma in cancer treatment. *Clinical Plasma Medicine* 1: 2–7. DOI: 10.1016/j.cpme.2013.08.001. 44

Schlüter O, Ehlbeck J, Hertel C, Habermeyer M, Roth A, Engel KH, Holzhauser T, Knorr D, Eisenbrand G (2013) Opinion on the use of plasma processes for treatment of foods. *Molecular Nutrition and Food Research* 57: 920–927. DOI: 10.1002/mnfr.201300039. 76

Schlüter O, Fröhling A (2014) Non–thermal processing | Cold plasma for bioefficient food processing. In: *Encyclopedia of Food Microbiology* (Second Edition), Batt C.A., Tortorello M.L. (eds.) pp. 948–953. Academic Press: Oxford. DOI: 10.1016/B978-0-12-384730-0.00402-X. 37

Schmidt A, Bekeschus S, Wende K, Vollmar B, von Woedtke T (2017a) A cold plasma jet accelerates wound healing in a murine model of full-thickness skin wounds. *Experimental Dermatology* 26: 156–162. DOI: 10.1111/exd.13156. 39

Schmidt A, von Woedtke T, Vollmar B, Hasse S, Bekeschus S (2019) Nrf2 signaling and inflammation are key events in physical plasma–spurred wound healing. *Theranostics* 9: 1066. DOI: 10.7150/thno.29754. 42, 43

Schmidt JB, Sands B, Scofield J, Gord JR, Roy S (2017b) Comparison of femtosecond–and nanosecond–two–photon–absorption laser–induced fluorescence (TALIF) of atomic oxygen in atmospheric–pressure plasmas. *Plasma Sources Science and Technology* 26: 055004. DOI: 10.1088/1361-6595/aa61be. 10

Schmidt–Bleker A, Norberg SA, Winter J, Johnsen E, Reuter S, Weltmann K, Kushner MJ (2015a) Propagation mechanisms of guided streamers in plasma jets: the influence of electronegativity of the surrounding gas. *Plasma Sources Science and Technology* 24: 035022. DOI: 10.1088/0963-0252/24/3/035022. 10

Schmidt–Bleker A, Reuter S, Weltmann K (2015b) Quantitative schlieren diagnostics for the determination of ambient species density, gas temperature and calorimetric power of cold atmospheric plasma jets. *Journal of Physics D: Applied Physics* 48: 175202. DOI: /10.1088/0022-3727/48/17/175202. 13

Schmidt–Bleker A, Reuter S, Weltmann K–D (2014) Non–dispersive path mapping approximation for the analysis of ambient species diffusion in laminar jets. *Physics of Fluids* 26: 083603. DOI: 10.1063/1.4893573. 12

Scholtz V, Pazlarova J, Souskova H, Khun J, Julak J (2015) Nonthermal plasma—A tool for decontamination and disinfection. *Biotechnology Advances* 33: 1108–1119. DOI: 10.1016/j.biotechadv.2015.01.002. 66, 73

Schönekerl S, Weigert A, Uhlig S, Wellner K, Pörschke R, Pfefferkorn C, Backhaus K, Lerch A (2020) Evaluating the performance of a lab–scale water treatment plant using non–thermal plasma technology. *Water* 12: 1956. DOI: 10.3390/w12071956. 69

Schuster M, Seebauer C, Rutkowski R, Hauschild A, Podmelle F, Metelmann C, Metelmann B, von Woedtke T, Hasse S, Weltmann K–D (2016) Visible tumor surface response to physical plasma and apoptotic cell kill in head and neck cancer. *Journal of Cranio–Maxillofacial Surgery* 44: 1445–1452. DOI: 10.1016/j.jcms.2016.07.001. 53

Schwarz F, Derks J, Monje A, Wang HL (2018) Peri-implantitis. *Journal of Clinical Periodontology* 45: S246–S266. DOI: 10.1111/jcpe.12954. 54

Schweitzer C, Schmidt R (2003) Physical mechanisms of generation and deactivation of singlet oxygen. *Chemical Reviews* 103: 1685–1758. DOI: 10.1021/cr010371d. 49

Seeger W (2005) Pneumologie–von der Phthisiologie zur regenerativen Medizin. *DMW–Deutsche Medizinische Wochenschrift* 130: 1543–1546. DOI: 10.1055/s-2005-870860. 54

Segat A, Misra N, Cullen P, Innocente N (2016) Effect of atmospheric pressure cold plasma (ACP) on activity and structure of alkaline phosphatase. *Food and Bioproducts Processing* 98: 181–188. DOI: 10.1016/j.fbp.2016.01.010. 32

Segui Y, Despax B, Raynaud P, Massines F (1996) Plasma research at the Electrical Engineering Laboratory of Paul Sabatier University–Toulouse–France. *Plasmas and Polymers* 1: 195–205. DOI: 10.1007/BF02532826. 3

Seinfeld JH, Pandis SN (2016) *Atmospheric Chemistry and Physics: From Air Pollution to Climate Change.* John Wiley and Sons. 18, 19

Selcuk M, Oksuz L, Basaran P (2008) Decontamination of grains and legumes infected with Aspergillus spp. and Penicillum spp. by cold plasma treatment. *Bioresource Technology* 99: 5104–5109. DOI: 10.1016/j.biortech.2007.09.076. 24, 28

Semmler ML, Bekeschus S, Schäfer M, Bernhardt T, Fischer T, Witzke K, Seebauer C, Rebl H, Grambow E, Vollmar B (2020) Molecular mechanisms of the efficacy of cold atmospheric pressure plasma (cap) in cancer treatment. *Cancers* 12: 269. DOI: 10.3390/cancers12020269. 48, 52

Sen CK, Gordillo GM, Roy S, Kirsner R, Lambert L, Hunt TK, Gottrup F, Gurtner GC, Longaker MT (2009) Human skin wounds: a major and snowballing threat to public health and the economy. *Wound Repair and Regeneration* 17: 763–771. DOI: 10.1111/j.1524-475X.2009.00543.x. 39

Service RF (2004) Newcomer heats up the race for practical fuel cells. American Association for the Advancement of Science. DOI: 10.1126/science.303.5654.29. 79

Setlow P (2007) I will survive: DNA protection in bacterial spores. *Trends in Microbiology* 15: 172–180. DOI: 10.1016/j.tim.2007.02.004. 37

Setsuhara Y (2016) Low–temperature atmospheric–pressure plasma sources for *Plasma Medicine*. *Archives of Biochemistry and Biophysics* 605: 3–10. DOI: 10.1016/j.abb.2016.04.009. 7

Shao P–L, Liao J–D, Wong T–W, Wang Y–C, Leu S, Yip H–K (2016) Enhancement of wound healing by non–thermal N2/Ar micro–plasma exposure in mice with fractional–CO_2–laser–induced wounds. *PloS One* 11: e0156699. DOI: 10.1371/journal.pone.0156699. 41

Shapira Y, Multanen V, Whyman G, Bormashenko Y, Chaniel G, Barkay Z, Bormashenko E (2017) Plasma treatment switches the regime of wetting and floating of pepper seeds. *Colloids and Surfaces B: Biointerfaces* 157: 417–423. DOI: 10.1016/j.colsurfb.2017.06.006. 24

Sharma P, Allison JP (2015) The future of immune checkpoint therapy. *Science* 348: 56–61. DOI: 10.1126/science.aaa8172. 52

Shashurin A, Keidar M (2015) Experimental approaches for studying non–equilibrium atmospheric plasma jets. *Physics of Plasmas* 22: 122002. DOI: 10.1063/1.4933365. 10, 11

Shashurin A, Shneider M, Dogariu A, Miles R, Keidar M (2010) Temporary–resolved measurement of electron density in small atmospheric plasmas. *Applied Physics Letters* 96: 171502. DOI: 10.1063/1.3389496. 11, 12

Shekargoftar M, Kelar J, Krumpolec R, Jurmanová J, Homola T (2018) A comparison of the effects of ambient air plasma generated by volume and by coplanar DBDs on the surfaces of PP/Al/PET laminated foil. *IEEE Transactions on Plasma Science* 46: 3653–3661. DOI: 10.1109/TPS.2018.2861085. 71

Shekhter A, Kabisov R, Pekshev A, Kozlov N, Perov YL (1998) Experimental and clinical validation of plasmadynamic therapy of wounds with nitric oxide. *Bulletin of Experimental Biology and Medicine* 126: 829–834. DOI: 10.1007/BF02446923. 41

Shekhter AB, Serezhenkov VA, Rudenko TG, Pekshev AV, Vanin AF (2005) Beneficial effect of gaseous nitric oxide on the healing of skin wounds. *Nitric Oxide* 12: 210–219. DOI: 10.1016/j.niox.2005.03.004. 41, 54

Shenton M, Stevens G (2001) Surface modification of polymer surfaces: atmospheric plasma versus vacuum plasma treatments. *Journal of Physics D: Applied Physics* 34: 2761. DOI: 10.1088/0022-3727/34/18/308. 3

Shi Q, Song K, Zhou X, Xiong Z, Du T, Lu X, Cao Y (2015) Effects of non-equilibrium plasma in the treatment of ligature-induced peri-implantitis. *Journal of Clinical Periodontology* 42: 478–487. DOI: 10.1111/jcpe.12403. 54

Shkurenkov I, Burnette D, Lempert WR, Adamovich IV (2014) Kinetics of excited states and radicals in a nanosecond pulse discharge and afterglow in nitrogen and air. *Plasma Sources Science and Technology* 23: 065003. DOI: 10.1088/0963-0252/23/6/065003. 16

Shneider MN, Miles RB (2005) Microwave diagnostics of small plasma objects. *Journal of Applied Physics* 98: 033301. DOI: 10.1063/1.1996835. 12

Shoemaker CA, Carlson WH (1990) pH affects seed germination of eight bedding plant species. *HortScience* 25: 762–764. DOI: 10.21273/HORTSCI.25.7.762. 28

Shohet JL (1991) Plasma–aided manufacturing. *IEEE Transactions on Plasma Science* 19: 725–733. DOI: 10.1109/27.108405. 3

Shulutko A, Antropova N, Kriuger I (2004) NO–therapy in the treatment of purulent and necrotic lesions of lower extremities in diabetic patients. *Khirurgiia*: 43–46. 54

Šimek M, Černák M, Kylián O, Foest R, Hegemann D, Martini R (2019) White paper on the future of plasma science for optics and glass. *Plasma Processes and Polymers* 16: 1700250. DOI: 10.1002/ppap.201700250. 77

Simoncelli E, Stancampiano A, Boselli M, Gherardi M, Colombo V (2019) Experimental investigation on the influence of target physical properties on an impinging plasma jet. *Plasma* 2: 369–379. DOI: 10.3390/plasma2030029. 9, 95

Šimončicová J, Kaliňáková B, Kováčik D, Medvecká V, Lakatoš B, Kryštofová S, Hoppanová L, Palušková V, Hudecová D, Ďurina P (2018) Cold plasma treatment triggers antioxidative defense system and induces changes in hyphal surface and subcellular structures of Aspergillus flavus. *Applied Microbiology and Biotechnology* 102: 6647–6658. DOI: 10.1007/s00253-018-9118-y. 24

Siniscalchi–Minna S, Bianchi FD, De–Prada–Gil M, Ocampo–Martinez C (2019) A wind farm control strategy for power reserve maximization. *Renewable Energy* 131: 37–44. DOI: 10.1016/j.renene.2018.06.112.

Sivachandiran L, Khacef A (2017) Enhanced seed germination and plant growth by atmospheric pressure cold air plasma: combined effect of seed and water treatment. *RSC Advances* 7: 1822–1832. DOI: 10.1039/C6RA24762H. 28, 29

Snoeckx R, Bogaerts A (2017) Plasma technology–a novel solution for CO 2 conversion? *Chemical Society Reviews* 46: 5805–5863. DOI: 10.1039/C6CS00066E. 6

Snoeckx R, Zeng Y, Tu X, Bogaerts A (2015) Plasma–based dry reforming: improving the conversion and energy efficiency in a dielectric barrier discharge. *RSC Advances* 5: 29799–29808. DOI: 10.1039/C5RA01100K. 71, 77

Soloveichik G (2019) Electrochemical synthesis of ammonia as a potential alternative to the Haber–Bosch process. *Nature Catalysis* 2: 377–380. DOI: 10.1038/s41929-019-0280-0. 87

Sonawane A, Mujawar MA, Bhansali S (2018) Cold atmospheric plasma annealing of plasmonic silver nanoparticles. *ECS Transactions* 88: 197. DOI: 10.1149/08801.0197ecst. 72

Sonawane A, Mujawar MA, Bhansali S (2019) Atmospheric plasma treatment enhances the biosensing properties of graphene oxide–silver nanoparticle Composite. *Journal of The Electrochemical Society* 166: B3084. DOI: 10.1149/2.0161909jes. 75, 76

Song HP, Kim B, Choe JH, Jung S, Moon SY, Choe W, Jo C (2009) Evaluation of atmospheric pressure plasma to improve the safety of sliced cheese and ham inoculated by 3–strain cocktail Listeria monocytogenes. *Food Microbiology* 26: 432–436. DOI: 10.1016/j.fm.2009.02.010. 30

Song Y, Fan X (2020) Cold plasma enhances the efficacy of aerosolized hydrogen peroxide in reducing populations of Salmonella Typhimurium and Listeria innocua on grape tomatoes, apples, cantaloupe and romaine lettuce. *Food Microbiology* 87: 103391. DOI: 10.1016/j.fm.2019.103391. 30

Sotiriou GA, Meyer A, Knijnenburg JT, Panke S, Pratsinis SE (2012) Quantifying the origin of released Ag^+ ions from nanosilver. *Langmuir* 28: 15929–15936. DOI: 10.1021/la303370d. 77

Staehelin J, Hoigne J (1982) Decomposition of ozone in water: rate of initiation by hydroxide ions and hydrogen peroxide. *Environmental Science and Technology* 16: 676–681. DOI: 10.1021/es00104a009. 19

Stafford DS, Kushner MJ (2004) O 2 (Δ 1) production in He/O 2 mixtures in flowing low pressure plasmas. *Journal of Applied Physics* 96: 2451–2465. DOI: 10.1063/1.1768615. 14, 15, 17

Stalder K, Vidmar R, Nersisyan G, Graham W (2006) Modeling the chemical kinetics of high–pressure glow discharges in mixtures of helium with real air. *Journal of Applied Physics* 99: 093301. DOI: 10.1063/1.2193170. 14

Stancampiano A, Chung T–H, Dozias S, Pouvesle J–M, Mir L, Robert É (2019) Mimicking of human body electrical characteristic for easier translation of plasma biomedical studies to clinical applications. *IEEE Transactions on Radiation and Plasma Medical Sciences* 4: 335–342. DOI: 10.1109/TRPMS.2019.2936667. 35

Steele Brian C, Heinzel A (2001) Materials for fuel–cell technologies. *Nature* 414: 345. DOI: 10.1038/35104620. 79

Štěpánová V, Slavíček P, Kelar J, Prášil J, Smékal M, Stupavská M, Jurmanová J, Černák M (2018) Atmospheric pressure plasma treatment of agricultural seeds of cucumber (Cucumis sativus L.) and pepper (Capsicum annuum L.) with effect on reduction of diseases and germination improvement. *Plasma Processes and Polymers* 15: 1700076. DOI: 10.1002/ppap.201700076. 27

Stere C, Adress W, Burch R, Chansai S, Goguet A, Graham W, Hardacre C (2015) Probing a non–thermal plasma activated heterogeneously catalyzed reaction using in situ DRIFTS–MS. *ACSCatalysis* 5: 956–964. DOI: 10.1021/cs5019265. 88

Stere CE, Anderson JA, Chansai S, Delgado JJ, Goguet A, Graham WG, Hardacre C, Taylor SR, Tu X, Wang Z (2017) Non-thermal plasma activation of gold-based catalysts for low-temperature water–gas shift catalysis. *Angewandte Chemie* 129: 5671–5675. DOI: 10.1002/ange.201612370. 87

Sternberg A, Bardow A (2015) Power–to–What?–Environmental assessment of energy storage systems. *Energy and Environmental Science* 8: 389–400. DOI: 10.1039/C4EE03051F. 95

Stevefelt J, Pouvesle J, Bouchoule A (1982) Reaction kinetics of a high pressure helium fast discharge afterglow. *The Journal of Chemical Physics* 76: 4006–4015. DOI: 10.1063/1.443521. 14

Stewig C, Schüttler S, Urbanietz T, Böke M, von Keudell A (2020) Excitation and dissociation of CO_2 heavily diluted in noble gas atmospheric pressure plasma. *Journal of Physics D: Applied Physics* 53: 125205. DOI: 10.1088/1361-6463/ab634f. 96

Stoffels E, Gonzalvo YA, Whitmore T, Seymour D, Rees J (2006) A plasma needle generates nitric oxide. *Plasma Sources Science and Technology* 15: 501. DOI: 10.1088/0963-0252/15/3/028. 40

Stoffels E, Sakiyama Y, Graves DB (2008) Cold atmospheric plasma: charged species and their interactions with cells and tissues. *IEEE Transactions on Plasma Science* 36: 1441–1457. DOI: 10.1109/TPS.2008.2001084. 35

Stougie L, Giustozzi N, van der Kooi H, Stoppato A (2018) Environmental, economic and ex-ergetic sustainability assessment of power generation from fossil and renewable energy sources. *International Journal of Energy Research* 42: 2916–2926. DOI: 10.1002/er.4037. 95

Subrahmanyam C, Renken A, Kiwi–Minsker L (2010) Catalytic non–thermal plasma reactor for abatement of toluene. *Chemical Engineering Journal* 160: 677–682. DOI: 10.1016/j.cej.2010.04.011. 61, 90

Suganya A, Shanmugavelayutham G, Rodríguez CS (2016) Study on structural, morphological and thermal properties of surface modified polyvinylchloride (PVC) film under air, argon and oxygen discharge plasma. *Materials Research Express* 3: 095302. DOI: 10.1088/2053-1591/3/9/095302. 80

Sultana S, Vandenbroucke AM, Leys C, De Geyter N, Morent R (2015) Abatement of VOCs with alternate adsorption and plasma–assisted regeneration: a review. *Catalysts* 5: 718–746. DOI: 10.3390/catal5020718. 93

Sun D, Hong R, Liu J, Wang F, Wang Y (2016) Preparation of carbon nanomaterials using two–group arc discharge plasma. *Chemical Engineering Journal* 303: 217–230. DOI: 10.1016/j.cej.2016.05.098. 80

Sun D–W (2014) *Emerging Technologies for Food Processing*. Elsevier. 30

Sun J, Madan R, Karp CL, Braciale TJ (2009) Effector T cells control lung inflammation during acute influenza virus infection by producing IL–10. *Nature Medicine* 15: 277–284. DOI: 10.1038/nm.1929. 56

Sun W, Uddi M, Won SH, Ombrello T, Carter C, Ju Y (2012a) Kinetic effects of non–equilibrium plasma–assisted methane oxidation on diffusion flame extinction limits. *Combustion and Flame* 159: 221–229. DOI: 10.1016/j.combustflame.2011.07.008. 92

Sun Y, Zhou L, Zhang L, Sui H (2012b) Synergistic effects of non–thermal plasma–assisted catalyst and ultrasound on toluene removal. *Journal of Environmental Sciences* 24: 891–896. DOI: 10.1016/S1001-0742(11)60842-5. 61

Surowsky B, Fischer A, Schlueter O, Knorr D (2013) Cold plasma effects on enzyme activity in a model food system. *Innovative Food Science and Emerging Technologies* 19: 146–152. DOI: 10.1016/j.ifset.2013.04.002. 30

Svarnas P, Giannakopoulos E, Kalavrouziotis I, Krontiras C, Georga S, Pasolari R, Papadopoulos P, Apostolou I, Chrysochoou D (2020) Sanitary effect of FE–DBD cold plasma in ambient air on sewage biosolids. *Science of the Total Environment* 705: 135940. DOI: 10.1016/j.scitotenv.2019.135940. 63

Szili EJ, Hong S–H, Oh J–S, Gaur N, Short RD (2018) Tracking the penetration of plasma reactive species in tissue models. *Trends in Biotechnology* 36: 594–602. DOI: 10.1016/j.tibtech.2017.07.012. 5, 35, 104, 107

Takamatsu T, Uehara K, Sasaki Y, Miyahara H, Matsumura Y, Iwasawa A, Ito N, Azuma T, Kohno M, Okino A (2014) Investigation of reactive species using various gas plasmas. *RSC Advances* 4: 39901–39905. DOI: 10.1039/C4RA05936K. 5

Talebizadeh P, Babaie M, Brown R, Rahimzadeh H, Ristovski Z, Arai M (2014) The role of non–thermal plasma technique in NO_x treatment: A review. *Renewable and Sustainable Energy Reviews* 40: 886–901. DOI: 10.1016/j.rser.2014.07.194. 62

Tanaka H, Ishikawa K, Mizuno M, Toyokuni S, Kajiyama H, Kikkawa F, Metelmann H–R, Hori M (2017) State of the art in medical applications using non–thermal atmospheric pressure plasma. *Reviews of Modern Plasma Physics* 1: 3. DOI: 10.1007/s41614-017-0004-3. 35

Tanaka K, Babic I, Nathanson D, Akhavan D, Guo D, Gini B, Dang J, Zhu S, Yang H, De Jesus J (2011) Oncogenic EGFR signaling activates an mTORC$_2$–NF–κB pathway that promotes chemotherapy resistance. *Cancer Discovery* 1: 524–538. DOI: 10.1158/2159-8290. CD-11-0124. 47

Tang X, Li K, Yi H, Ning P, Xiang Y, Wang J, Wang C (2012) MnOx catalysts modified by nonthermal plasma for NO catalytic oxidation. *The Journal of Physical Chemistry* C 116: 10017–10028. DOI: 10.1021/jp300664f. 84

Tappi S, Gozzi G, Vannini L, Berardinelli A, Romani S, Ragni L, Rocculi P (2016) Cold plasma treatment for fresh–cut melon stabilization. *Innovative Food Science and Emerging Technologies* 33: 225–233. DOI: 10.1016/j.ifset.2015.12.022. 31

Taranto J, Frochot D, Pichat P (2007) Combining cold plasma and TiO2 photocatalysis to purify gaseous effluents: a preliminary study using methanol–contaminated air. *Industrial and Engineering Chemistry Research* 46: 7611–7614. DOI: 10.1021/ie0700967. 104

Ten Bosch L, Habedank B, Siebert D, Mrotzek J, Viöl W (2019) Cold atmospheric pressure plasma comb—A physical approach for pediculosis treatment. *International Journal of Environmental Research and Public Health* 16: 19. DOI: 10.3390/ijerph16010019. 35

Tendero C, Tixier C, Tristant P, Desmaison J, Leprince P (2006) Atmospheric pressure plasmas: A review. *Spectrochimica Acta Part B: Atomic Spectroscopy* 61: 2–30. DOI: 10.1016/j. sab.2005.10.003. 3, 5, 8, 77

Thepsithar P, Roberts EP (2006) Removal of phenol from contaminated kaolin using electrokinetically enhanced in situ chemical oxidation. *Environmental Science and Technology* 40: 6098–6103. DOI: 10.1021/es060883f. 66

Thévenet F, Sivachandiran L, Guaitella O, Barakat C, Rousseau A (2014) Plasma–catalyst coupling for volatile organic compound removal and indoor air treatment: a review. *Journal of Physics D: Applied Physics* 47: 224011. DOI: 10.1088/0022-3727/47/22/224011. 59

Thirumdas R, Trimukhe A, Deshmukh R, Annapure U (2017) Functional and rheological properties of cold plasma treated rice starch. *Carbohydrate Polymers* 157: 1723–1731. DOI: 10.1016/j.carbpol.2016.11.050. 31

Thuan NT, Chang MB (2012) Investigation of the degradation of pentachlorophenol in sandy soil via low–temperature pyrolysis. *Journal of Hazardous Materials* 229: 411–418. DOI: 10.1016/j.jhazmat.2012.06.027. 66

Tian W, Kushner MJ (2014) Atmospheric pressure dielectric barrier discharges interacting with liquid covered tissue. *Journal of Physics D: Applied Physics* 47: 165201. DOI: 10.1088/0022-3727/47/16/165201. 21, 44, 49

Tichonovas M, Krugly E, Racys V, Hippler R, Kauneliene V, Stasiulaitiene I, Martuzevicius D (2013) Degradation of various textile dyes as wastewater pollutants under dielectric barrier discharge plasma treatment. *Chemical Engineering Journal* 229: 9–19. DOI: 10.1016/j.cej.2013.05.095. 64

Tiya–Djowe A, Acayanka E, Mbouopda AP, Boyom–Tatchemo W, Laminsi S, Gaigneaux EM (2019) Producing oxide catalysts by exploiting the chemistry of gliding arc atmospheric plasma in humid air. *Catalysis Today* 334: 104–112. DOI: 10.1016/j.cattod.2019.01.008. 71, 74, 80

Tiya–Djowe A, Laminsi S, Noupeyi GL, Gaigneaux EM (2015) Non–thermal plasma synthesis of sea–urchin like α–FeOOH for the catalytic oxidation of Orange II in aqueous solution. *Applied Catalysis B: Environmental* 176: 99–106. DOI: 10.1016/j.apcatb.2015.03.053. 85

Tolouie H, Mohammadifar MA, Ghomi H, Hashemi M (2018) Cold atmospheric plasma manipulation of proteins in food systems. *Critical Reviews in Food Science and Nutrition* 58: 2583–2597. DOI: 10.1080/10408398.2017.1335689. 8

Trelles JP (2018) Advances and challenges in computational fluid dynamics of atmospheric pressure plasmas. *Plasma Sources Science and Technology* 27: 093001. DOI: 10.1088/1361-6595/aac9fa. 13

Uddi M, Jiang N, Adamovich I, Lempert W (2009) Nitric oxide density measurements in air and air/fuel nanosecond pulse discharges by laser induced fluorescence. *Journal of Physics D: Applied Physics* 42: 075205. DOI: 10.1088/0022-3727/42/7/075205. 16

Ulrich C, Kluschke F, Patzelt A, Vandersee S, Czaika V, Richter H, Bob A, Hutten Jv, Painsi C, Hüge R (2015) Clinical use of cold atmospheric pressure argon plasma in chronic leg ulcers: A pilot study. *Journal of Wound Care* 24: 196–203. DOI: 10.12968/jowc.2015.24.5.196. 42

Ulu M, Pekbagriyanik T, Ibis F, Enhos S, Ercan U (2018) Antibiofilm efficacies of cold plasma and Er: YAG laser on Staphylococcus aureus biofilm on titanium for nonsurgical treatment of peri-implantitis. *Nigerian Journal of Clinical Practice* 21: 758–765. DOI: 10.4103/njcp.njcp_261_17. 54

Urashima K, Chang J–S (2000) Removal of volatile organic compounds from air streams and industrial flue gases by non–thermal plasma technology. *IEEE Transactions on Dielectrics and Electrical Insulation* 7: 602–614. DOI: 10.1109/94.879356. 90

Utsumi F, Kajiyama H, Nakamura K, Tanaka H, Mizuno M, Ishikawa K, Kondo H, Kano H, Hori M, Kikkawa F (2013) Effect of indirect nonequilibrium atmospheric pressure plasma on anti–proliferative activity against chronic chemo–resistant ovarian cancer cells *in vitro* and *in vivo*. *PloS One* 8: e81576. DOI: 10.1371/journal.pone.0081576. 46

Van der Ham CJ, Koper MT, Hetterscheid DG (2014) Challenges in reduction of dinitrogen by proton and electron transfer. *Chemical Society Reviews* 43: 5183–5191. DOI: 10.1039/C4CS00085D. 77

Van Deynse A, Morent R, De Geyter N (2016) Surface modification of polymers using atmospheric pressure cold plasma technology. In: *Polymer Science: Research Advances, Pratical Applications and Educational Aspects*, pp. 506–516. Formatex Research Center. 104

Van Doremaele E, Kondeti V, Bruggeman PJ (2018) Effect of plasma on gas flow and air concentration in the effluent of a pulsed cold atmospheric pressure helium plasma jet. *Plasma Sources Science and Technology* 27: 095006. DOI: 10.1088/1361-6595/aadbd3. 9

Van Durme J, Dewulf J, Leys C, Van Langenhove H (2008) Combining non–thermal plasma with heterogeneous catalysis in waste gas treatment: A review. *Applied Catalysis B: Environmental* 78: 324–333. DOI: 10.1016/j.apcatb.2007.09.035. 89

Van Gaens W, Bogaerts A (2013) Kinetic modelling for an atmospheric pressure argon plasma jet in humid air. *Journal of Physics D: Applied Physics* 46: 275201. DOI: 10.1088/0022-3727/46/27/275201. 5, 16

Van Gils C, Hofmann S, Boekema B, Brandenburg R, Bruggeman P (2013) Mechanisms of bacterial inactivation in the liquid phase induced by a remote RF cold atmospheric pressure plasma jet. *Journal of Physics D: Applied Physics* 46: 175203. DOI: 10.1088/0022-3727/46/17/175203. 28

van Koppen CJ, Hartmann RW (2015) Advances in the treatment of chronic wounds: a patent review. *Expert Opinion on Therapeutic Patents* 25: 931–937. DOI: 10.1517/13543776.2015.1045879. 39

Van Nguyen D, Ho NM, Hoang KD, Le TV, Le VH (2020) An investigation on treatment of groundwater with cold plasma for domestic water supply. *Groundwater for Sustainable Development* 10: 100309. DOI: 10.1016/j.gsd.2019.100309. 63

Vandenbroucke AM, Dinh MN, Nuns N, Giraudon J–M, De Geyter N, Leys C, Lamonier J–F, Morent R (2016) Combination of non–thermal plasma and Pd/LaMnO$_3$ for dilute trichloroethylene abatement. *Chemical Engineering Journal* 283: 668–675. DOI: 10.1016/j.cej.2015.07.089. 89, 92

Vandenbroucke AM, Morent R, De Geyter N, Leys C (2011) Non–thermal plasmas for non–catalytic and catalytic VOC abatement. *Journal of Hazardous Materials* 195: 30–54. DOI: 10.1016/j.jhazmat.2011.08.060. 59

Vanraes P, Bogaerts A (2018) Plasma physics of liquids—A focused review. *Applied Physics Reviews* 5: 031103. DOI: 10.1063/1.5020511. 28, 103

Velusamy T, Liguori A, Macias-Montero M, Padmanaban DB, Carolan D, Gherardi M, Colombo V, Maguire P, Svrcek V, Mariotti D (2017) Ultra-small CuO nanoparticles with tailored energy-band diagram synthesized by a hybrid plasma-liquid process. *Plasma Processes and Polymers* 14: 1600224. DOI: 10.1002/ppap.201600224. 80

Vesel A, Mozetic M (2017) New developments in surface functionalization of polymers using controlled plasma treatments. *Journal of Physics D: Applied Physics* 50: 293001. DOI: 10.1088/1361-6463/aa748a. 6

Vladimirov SV, Ostrikov K (2004) Dynamic self–organization phenomena in complex ionized gas systems: new paradigms and technological aspects. *Physics Reports* 393: 175–380. DOI: 10.1016/j.physrep.2003.12.003. 9

Vogelsang A, Ohl A, Steffen H, Foest R, Schröder K, Weltmann KD (2010) Locally resolved analysis of polymer surface functionalization by an atmospheric pressure argon microplasma jet with air entrainment. *Plasma Processes and Polymers* 7: 16–24. DOI: 10.1002/ppap.200900091. 54

Vojvodic A, Medford AJ, Studt F, Abild–Pedersen F, Khan TS, Bligaard T, Nørskov J (2014) Exploring the limits: A low–pressure, low–temperature Haber–Bosch process. *Chemical Physics Letters* 598: 108–112. DOI: 10.1016/j.cplett.2014.03.003. 87

Volin JC, Denes FS, Young RA, Park SM (2000) Modification of seed germination performance through cold plasma chemistry technology. *Crop Science* 40: 1706–1718. DOI: 10.2135/cropsci2000.4061706x. 24

Volotskova O, Hawley TS, Stepp MA, Keidar M (2012) Targeting the cancer cell cycle by cold atmospheric plasma. *Scientific Reports* 2: 636. DOI: 10.1038/srep00636. 46

Von Woedtke T, Reuter S, Masur K, Weltmann K–D (2013) Plasmas for medicine. *Physics Reports* 530: 291–320. DOI: 10.1016/j.physrep.2013.05.005. 35, 103, 104

von Woedtke T, Schmidt A, Bekeschus S, Wende K, Weltmann K–D (2019) Plasma Medicine: a field of applied redox biology. *in vivo* 33: 1011–1026. DOI: 10.21873/invivo.11570. 35, 42

Vujošević D, Mozetič M, Cvelbar U, Krstulović N, Milošević S (2007) Optical emission spectroscopy characterization of oxygen plasma during degradation of Escherichia coli. *Journal of Applied Physics* 101: 103305. DOI: 10.1063/1.2732693. 73

Wakabayashi N, Slocum SL, Skoko JJ, Shin S, Kensler TW (2010) When NRF2 talks, who's listening? *Antioxidants and Redox Signaling* 13: 1649–1663. DOI: 10.1089/ars.2010.3216. 42

Wallenhorst L, Gurău L, Gellerich A, Militz H, Ohms G, Viöl W (2018) UV–blocking properties of Zn/ZnO coatings on wood deposited by cold plasma spraying at atmospheric pressure. *Applied Surface Science* 434: 1183–1192. DOI: 10.1016/j.apsusc.2017.10.214. 71

Walsh JL, Kong MG (2008) Contrasting characteristics of linear–field and cross–field atmospheric plasma jets. *Applied Physics Letters* 93: 111501. DOI: 10.1063/1.2982497. 49

Wang B, Chi C, Xu M, Wang C, Meng D (2017a) Plasma–catalytic removal of toluene over CeO_2–MnO_x catalysts in an atmosphere dielectric barrier discharge. *Chemical Engineering Journal* 322: 679–692. DOI: 10.1016/j.cej.2017.03.153. 87, 88

Wang J, Guo Y, Gao J, Jin X, Wang Z, Wang B, Li K, Li Y (2011) Detection and comparison of reactive oxygen species (ROS) generated by chlorophyllin metal (Fe, Mg and Cu) complexes under ultrasonic and visible–light irradiation. *Ultrasonics Sonochemistry* 18: 1028–1034. DOI: 10.1016/j.ultsonch.2010.12.006. 52

Wang L, Yi Y, Guo H, Tu X (2018a) Atmospheric pressure and room temperature synthesis of methanol through plasma–catalytic hydrogenation of CO_2. *ACS Catalysis* 8: 90–100. DOI: 10.1021/acscatal.7b02733. 87, 88

Wang M, Favi P, Cheng X, Golshan NH, Ziemer KS, Keidar M, Webster TJ (2016a) Cold atmospheric plasma (CAP) surface nanomodified 3D printed polylactic acid (PLA) scaffolds for bone regeneration. *Acta Biomaterialia* 46: 256–265. DOI: 10.1016/j.actbio.2016.09.030. 74, 75, 81

Wang T, Ren J, Qu G, Liang D, Hu S (2016b) Glyphosate contaminated soil remediation by atmospheric pressure dielectric barrier discharge plasma and its residual toxicity evaluation. *Journal of Hazardous Materials* 320: 539–546. DOI: 10.1016/j.jhazmat.2016.08.067. 67, 68

Wang T, Zhang X, Liu J, Liu H, Wang Y, Sun B (2018b) Effects of temperature on NOx removal with Mn–Cu/ZSM5 catalysts assisted by plasma. *Applied Thermal Engineering* 130: 1224–1232. DOI: 10.1016/j.applthermaleng.2017.11.113. 84

Wang TC, Lu N, Li J, Wu Y (2010) Evaluation of the potential of pentachlorophenol degradation in soil by pulsed corona discharge plasma from soil characteristics. *Environmental Science and Technology* 44: 3105–3110. DOI: 10.1021/es903527w. 67

Wang TC, Qu G, Li J, Liang D (2014a) Evaluation of the potential of soil remediation by direct multi–channel pulsed corona discharge in soil. *Journal of Hazardous Materials* 264: 169–175. DOI: 10.1016/j.jhazmat.2013.11.011. 67

Wang W, He J, Wu S (2020a) The definition and risks of cytokine release syndrome–like in 11 COVID–19–infected pneumonia critically ill patients: disease characteristics and retrospective analysis. *Medrxiv.* 56

Wang W, Wang H, Zhu T, Fan X (2015) Removal of gas phase low–concentration toluene over Mn, Ag and Ce modified HZSM–5 catalysts by periodical operation of adsorption and non–thermal plasma regeneration. *Journal of Hazardous Materials* 292: 70–78. DOI: 10.1016/j.jhazmat.2015.03.013. 61, 86

Wang X, Lin J–p, Min J–y, Wang P–C, Sun C–c (2018c) Effect of atmospheric pressure plasma treatment on strength of adhesive–bonded aluminum AA5052. *The Journal of Adhesion* 94: 701–722. DOI: 10.1080/00218464.2017.1393747. 72

Wang X, Zheng Y, Xu Z, Liu Y, Wang X (2014b) Low–temperature NO reduction with NH 3 over Mn–CeO x/CNT catalysts prepared by a liquid–phase method. *Catalysis Science and Technology* 4: 1738–1741. DOI: 10.1039/C4CY00026A. 87

Wang X–Q, Zhou R–W, de Groot G, Bazaka K, Murphy AB, Ostrikov KK (2017b) Spectral characteristics of cotton seeds treated by a dielectric barrier discharge plasma. *Scientific Reports* 7: 1–9. DOI: 10.1038/s41598-017-04963-4. 24

Wang Y, Wang Z, Zhu X, Yuan Y, Gao Z, Yue T (2020b) Application of electrical discharge plasma on the inactivation of Zygosaccharomyces rouxii in apple juice. *LWT* 121: 108974. DOI: 10.1016/j.lwt.2019.108974. 23

Warning A, Datta AK (2013) Interdisciplinary engineering approaches to study how pathogenic bacteria interact with fresh produce. *Journal of Food Engineering* 114: 426–448. DOI: 10.1016/j.jfoodeng.2012.09.004. 30

Waskow A, Betschart J, Butscher D, Oberbossel G, Klöti D, Büttner–Mainik A, Adamcik J, von Rohr PR, Schuppler M (2018) Characterization of efficiency and mechanisms of cold atmospheric pressure plasma decontamination of seeds for sprout production. *Frontiers in Microbiology* 9: 3164. DOI: 10.3389/fmicb.2018.03164. 24

Wegierski T, Steffl D, Kopp C, Tauber R, Buchholz B, Nitschke R, Kuehn EW, Walz G, Köttgen M (2009) TRPP2 channels regulate apoptosis through the Ca^{2+} concentration in the endoplasmic reticulum. *The EMBO Journal* 28: 490–499. DOI: 10.1038/emboj.2008.307. 50

Weinan E, Engquist B, Li X, Ren W, Vanden–Eijnden E (2007) Heterogeneous multiscale methods: a review. *Communications in Computational Physics* 2: 367–450. 14

Weiss M, Stope MB (2018) Physical plasma: a new treatment option in gynecological oncology. *Archives of Gynecology and Obstetrics* 298: 853–855. DOI: 10.1007/s00404-018-4889-z. 53

Weissker U, Hampel S, Leonhardt A, Büchner B (2010) Carbon nanotubes filled with ferromagnetic materials. *Materials* 3: 4387–4427. DOI: 10.3390/ma3084387. 80

Weltmann K–D, Von Woedtke T (2011) Basic requirements for plasma sources in medicine. *The European Physical Journal–Applied Physics* 55. DOI: 10.1051/epjap/2011100452. 4

Weltmann KD, Kindel E, Brandenburg R, Meyer C, Bussiahn R, Wilke C, Von Woedtke T (2009) Atmospheric pressure plasma jet for medical therapy: plasma parameters and risk estimation. *Contributions to Plasma Physics* 49: 631–640. DOI: 10.1002/ctpp.200910067. 39, 57

Weltmann KD, Kolb JF, Holub M, Uhrlandt D, Šimek M, Ostrikov K, Hamaguchi S, Cvelbar U, Černák M, Locke B (2019) The future for plasma science and technology. *Plasma Processes and Polymers* 16: 1800118. DOI: 10.1002/ppap.201800118. 81

Wen Q, Zhou W, Wang Y, Qing Y, Luo F, Zhu D, Huang Z (2017) Enhanced microwave absorption of plasma–sprayed Ti_3SiC_2/glass composite coatings. *Journal of Materials Science* 52: 832–842. DOI: 10.1007/s10853-016-0379-5. 71

Wende K, von Woedtke T, Weltmann K–D, Bekeschus S (2018) Chemistry and biochemistry of cold physical plasma derived reactive species in liquids. *Biological Chemistry* 400: 19–38. DOI: 10.1515/hsz-2018-0242. 6, 28

Whitehead JC (2010) Plasma catalysis: A solution for environmental problems. *Pure and Applied Chemistry* 82: 1329–1336. DOI: 10.1351/PAC-CON-10-02-39. 59

Whitehead JC (2016) Plasma–catalysis: the known knowns, the known unknowns and the unknown unknowns. *Journal of Physics D: Applied Physics* 49: 243001. DOI: 10.1088/0022-3727/49/24/243001. 88

Winter J, Brandenburg R, Weltmann K (2015) Atmospheric pressure plasma jets: an overview of devices and new directions. *Plasma Sources Science and Technology* 24: 064001. DOI: 10.1088/0963-0252/24/6/064001. 5

Wirtz M, Stoffels I, Dissemond J, Schadendorf D, Roesch A (2018) Actinic keratoses treated with cold atmospheric plasma. *Journal of the European Academy of Dermatology and Venereology* 32: e37–e39. DOI: 10.1111/jdv.14465. 53

Witzke M, Rumbach P, Go DB, Sankaran RM (2012) Evidence for the electrolysis of water by atmospheric–pressure plasmas formed at the surface of aqueous solutions. *Journal of Physics D: Applied Physics* 45: 442001. DOI: 10.1088/0022-3727/45/44/442001. 77

Wolf R, Sparavigna AC (2010) Role of plasma surface treatments on wetting and adhesion. *Engineering* 2: 397. DOI: 10.4236/eng.2010.26052. 73

Wu X, Thi VLD, Huang Y, Billerbeck E, Saha D, Hoffmann H–H, Wang Y, Silva LAV, Sarbanes S, Sun T (2018) Intrinsic immunity shapes viral resistance of stem cells. *Cell* 172: 423–438. e425. DOI: 10.1016/j.cell.2017.11.018. 56

Wu Z, McGoogan JM (2020) Characteristics of and important lessons from the coronavirus disease 2019 (COVID–19) outbreak in China: summary of a report of 72,314 cases from the Chinese Center for Disease Control and Prevention. *JAMA* 323: 1239–1242. DOI: 10.1001/jama.2020.2648. 37

Xia J, Zeng W, Xia Y, Wang B, Xu D, Liu D, Kong MG, Dong Y (2019) Cold atmospheric plasma induces apoptosis of melanoma cells via Sestrin2-mediated nitric oxide synthase signaling. *Journal of Biophotonics* 12: e201800046. DOI: 10.1002/jbio.201800046. 49

Xian Y, Zhang P, Lu X, Pei X, Wu S, Xiong Q, Ostrikov KK (2013) From short pulses to short breaks: Exotic plasma bullets via residual electron control. *Scientific Reports* 3: 1599. DOI: 10.1038/srep01599. 6

Xiang L, Xu X, Zhang S, Cai D, Dai X (2018) Cold atmospheric plasma conveys selectivity on triple negative breast cancer cells both *in vitro* and *in vivo*. *Free Radical Biology and Medicine* 124: 205–213. DOI: 10.1016/j.freeradbiomed.2018.06.001. 46, 47

Xie Q–w, Cho HJ, Calaycay J, Mumford RA, Swiderek KM, Lee TD, Ding A, Troso T, Nathan C (1992) Cloning and characterization of inducible nitric oxide synthase from mouse macrophages. *Science* 256: 225–228. DOI: 10.1126/science.1373522. 41

Xu C–H, Chiu Y–F, Yeh P–W, Chen J–Z (2016a) SnO2/CNT nanocomposite supercapacitors fabricated using scanning atmospheric–pressure plasma jets. *Materials Research Express* 3: 085002. DOI: 10.1088/2053-1591/3/8/085002. 71, 98

Xu C–H, Shen P–Y, Chiu Y–F, Yeh P–W, Chen C–C, Chen L–C, Hsu C–C, Cheng I–C, Chen J–Z (2016b) Atmospheric pressure plasma jet processed nanoporous Fe2O3/CNT composites for supercapacitor application. *Journal of Alloys and Compounds* 676: 469–473. DOI: 10.1016/j.jallcom.2016.03.185. 71, 98

Xu D, Xu Y, Cui Q, Liu D, Liu Z, Wang X, Yang Y, Feng M, Liang R, Chen H (2018a) Cold atmospheric plasma as a potential tool for multiple myeloma treatment. *Oncotarget* 9: 18002. DOI: 10.18632/oncotarget.24649. 46

Xu GM, Shi XM, Cai JF, Chen SL, Li P, Yao CW, Chang ZS, Zhang GJ (2015) Dual effects of atmospheric pressure plasma jet on skin wound healing of mice. *Wound Repair and Regeneration* 23: 878–884. DOI: 10.1111/wrr.12364. 41

Xu H, Ma R, Zhu Y, Du M, Zhang H, Jiao Z (2020) A systematic study of the antimicrobial mechanisms of cold atmospheric–pressure plasma for water disinfection. *Science of the Total Environment* 703: 134965. DOI: 10.1016/j.scitotenv.2019.134965. 63

Xu R, Wang S, Mei K (2007) Experimental study of low temperature plasma and catalyze process on formaldehyde decomposition. *Gaodianya Jishu/ High Voltage Engineering* 33: 178–181. 62

Xu R–G, Chen Z, Keidar M, Leng Y (2018b) The impact of radicals in cold atmospheric plasma on the structural modification of gap junction: A reactive molecular dynamics study. *International Journal of Smart and Nano Materials*. DOI: 10.1080/19475411.2018.1541936. 49, 51, 53, 56

Xu Y, Jin S, Xu H, Nagai A, Jiang D (2013) Conjugated microporous polymers: design, synthesis and application. *Chemical Society Reviews* 42: 8012–8031. DOI: 10.1039/c3cs60160a. 80

Xu Y, Li J, Tan Q, Peters AL, Yang C (2018c) Global status of recycling waste solar panels: A review. *Waste Management* 75: 450–458. DOI: 10.1016/j.wasman.2018.01.036. 95

Yadav IC, Devi NL (2017) Pesticides classification and its impact on human and environment. *Environmental Science and Engineering* 6: 140–158. 23

Yáñez–Pacios AJ, Martín–Martínez JM (2018) Comparative adhesion, ageing resistance, and surface properties of wood plastic composite treated with low pressure plasma and atmospheric pressure plasma jet. *Polymers* 10: 643. DOI: 10.3390/polym10060643. 71

Yang A, Wang X, Rong M, Liu D, Iza F, Kong MG (2011) 1–D fluid model of atmospheric–pressure rf $He+O_2$ cold plasmas: Parametric study and critical evaluation. *Physics of Plasmas* 18: 113503. DOI: 10.1063/1.3655441. 13

Yang J, De Guzman RC, Salley SO, Ng KS, Chen B–H, Cheng MM–C (2014) Plasma enhanced chemical vapor deposition silicon nitride for a high–performance lithium ion battery anode. *Journal of Power Sources* 269: 520–525. DOI: 10.1016/j.jpowsour.2014.06.135. 98

Yang J, Pu Y, Miao D, Ning X (2018a) Fabrication of durably superhydrophobic cotton fabrics by atmospheric pressure plasma treatment with a siloxane precursor. *Polymers* 10: 460. DOI: 10.3390/polym10040460. 71

Yang X, Kattel S, Senanayake SD, Boscoboinik JA, Nie X, Graciani Js, Rodriguez JA, Liu P, Stacchiola DJ, Chen JG (2015) Low pressure CO_2 hydrogenation to methanol over gold nanoparticles activated on a CeO x/TiO_2 interface. *Journal of the American Chemical Society* 137: 10104–10107. DOI: 10.1021/jacs.5b06150. 87

Yang Y, Guo J, Zhou X, Liu Z, Wang C, Wang K, Zhang J, Wang Z (2018b) A novel cold atmospheric pressure air plasma jet for peri–implantitis treatment: An *in vitro* study. *Dental Materials Journal* 37: 157–166. DOI: 10.4012/dmj.2017-030. 55

Yap TA, Sandhu SK, Carden CP, De Bono JS (2011) Poly (ADP-Ribose) polymerase (PARP) inhibitors: Exploiting a synthetic lethal strategy in the clinic. *CA: A Cancer Journal for Clinicians* 61: 31–49. 31 DOI: 10.3322/caac.20095. 44

Ye Y, Yu J, Wen D, Kahkoska AR, Gu Z (2018) Polymeric microneedles for transdermal protein delivery. *Advanced Drug Delivery Reviews* 127: 106–118. DOI: 10.1016/j.addr.2018.01.015. 52

Yehia S, Zarif M, Bita B, Teodorescu M, Carpen L, Vizireanu S, Petrea N, Dinescu G (2020) Development and optimization of single filament plasma jets for wastewater decontamination. *Plasma Chemistry and Plasma Processing*: 1–21. DOI: 10.1007/s11090-020-10111-0. 104

Yin Y, Yang T, Li Z, Liu P, Liu X, Devid EJ, Huang Q, Auerbach D, Kleyn AW (2019) Plasma driven boudouard reaction for efficient chemical storage. *SSRN Electronic Journal*. DOI: 10.2139/ssrn.3383798. 96

Yodpitak S, Mahatheeranont S, Boonyawan D, Sookwong P, Roytrakul S, Norkaew O (2019) Cold plasma treatment to improve germination and enhance the bioactive phytochemical content of germinated brown rice. *Food Chemistry* 289: 328–339. DOI: 10.1016/j.foodchem.2019.03.061. 103

Yu P, McKinnon JJ, Christensen CR, Christensen DA, Marinkovic NS, Miller LM (2003) Chemical imaging of microstructures of plant tissues within cellular dimension using synchrotron infrared microspectroscopy. *Journal of Agricultural and Food Chemistry* 51: 6062–6067. DOI: 10.1021/jf034654d. 27

Yuan S, Zheng Z, Chen J, Lu X (2009) Use of solar cell in electrokinetic remediation of cadmium–contaminated soil. *Journal of Hazardous Materials* 162: 1583–1587. DOI: 10.1016/j.jhazmat.2008.06.038. 66

Yusupov M, Bogaerts A, Huygh S, Snoeckx R, van Duin AC, Neyts EC (2013) Plasma–induced destruction of bacterial cell wall components: a reactive molecular dynamics simulation. *The Journal of Physical Chemistry* C 117: 5993–5998. DOI: /10.1021/jp3128516. 37

Yusupov M, Neyts E, Khalilov U, Snoeckx R, Van Duin A, Bogaerts A (2012) Atomic–scale simulations of reactive oxygen plasma species interacting with bacterial cell walls. *New Journal of Physics* 14: 093043. DOI: 10.1088/1367-2630/14/9/093043. 37

Zadi T, Azizi M, Nasrallah N, Bouzaza A, Maachi R, Wolbert D, Rtimi S, Assadi AA (2020) Indoor air treatment of refrigerated food chambers with synergetic association between cold plasma and photocatalysis: Process performance and photocatalytic poisoning. *Chemical Engineering Journal* 382: 122951. DOI: 10.1016/j.cej.2019.122951. 59

Zahoranová A, Hoppanová L, Šimončicová J, Tučeková Z, Medvecká V, Hudecová D, Kaliňáková B, Kováčik D, Černák M (2018) Effect of cold atmospheric pressure plasma on maize seeds: enhancement of seedlings growth and surface microorganisms inactivation. *Plasma Chemistry and Plasma Processing* 38: 969–988. DOI: 10.1007/s11090-018-9913-3. 24, 25

Zaitsev A, Poncin–Epaillard F, Lacoste A, Kassiba A, Debarnot D (2019) A multi–step cold plasma process for fine tuning of polymer nanostructuring. *Progress in Organic Coatings* 128: 112–119. DOI: 10.1016/j.porgcoat.2018.12.020. 81

Zarshenas K, Raisi A, Aroujalian A (2015) Surface modification of polyamide composite membranes by corona air plasma for gas separation applications. *RSC Advances* 5: 19760–19772. DOI: 10.1039/C4RA15547E. 6

Zeeshan HMA, Lee GH, Kim H–R, Chae H–J (2016) Endoplasmic reticulum stress and associated ROS. *International Journal of Molecular Sciences* 17: 327. DOI: 10.3390/ijms17030327. 50

Zen S, Abe T, Teramoto Y (2018) Indirect synthesis system for ammonia from nitrogen and water using nonthermal plasma under ambient conditions. *Plasma Chemistry and Plasma Processing* 38: 347–354. DOI: 10.1007/s11090-017-9869-8. 97

Zen S, Abe T, Teramoto Y (2019) Atmospheric pressure nonthermal plasma synthesis of magnesium nitride as a safe ammonia carrier. *Plasma Chemistry and Plasma Processing* 39: 1203–1210. DOI: 10.1007/s11090-019-10002-z. 96, 97

Zeng Y, Zhu X, Mei D, Ashford B, Tu X (2015) Plasma–catalytic dry reforming of methane over γ–Al_2O_3 supported metal catalysts. *Catalysis Today* 256: 80–87. DOI: 10.1016/j.cattod.2015.02.007. 92

Zhan J, Liu Y, Cheng W, Zhang A, Li R, Li X, Ognier S, Cai S, Yang C, Liu J (2018) Remediation of soil contaminated by fluorene using needle–plate pulsed corona discharge plasma. *Chemical Engineering Journal* 334: 2124–2133. DOI: 10.1016/j.cej.2017.11.093. 67, 68

Zhang H, Li X, Zhu F, Bo Z, Cen K, Tu X (2015a) Non–oxidative decomposition of methanol into hydrogen in a rotating gliding arc plasma reactor. *International Journal of Hydrogen Energy* 40: 15901–15912. DOI: 10.1016/j.ijhydene.2015.09.052. 71

Zhang H, Xu Z, Shen J, Li X, Ding L, Ma J, Lan Y, Xia W, Cheng C, Sun Q (2015b) Effects and mechanism of atmospheric–pressure dielectric barrier discharge cold plasma on lactate dehydrogenase (LDH) enzyme. *Scientific Reports* 5: 10031. DOI: 10.1038/srep10031. 31

Zhang H, Zhu F, Li X, Cen K, Du C, Tu X (2016) Enhanced hydrogen production by methanol decomposition using a novel rotating gliding arc discharge plasma. *RSC Advances* 6: 12770–12781. DOI: 10.1039/C5RA26343C. 71

Zhang R, Humphreys I, Sahu RP, Shi Y, Srivastava SK (2008) *In vitro* and *in vivo* induction of apoptosis by capsaicin in pancreatic cancer cells is mediated through ROS generation and mitochondrial death pathway. *Apoptosis* 13: 1465–1478. DOI: 10.1007/s10495-008-0278-6. 45

Zhang S, Li X–S, Zhu B, Liu J–L, Zhu X, Zhu A–M, Jang BW–L (2015c) Atmospheric–pressure O_2 plasma treatment of Au/TiO_2 catalysts for CO oxidation. *Catalysis Today* 256: 142–147. DOI: 10.1016/j.cattod.2015.04.027. 86

Zhang S–L, Liu T, Li C–J, Yao S–W, Li C–X, Yang G–J, Liu M (2015d) Atmospheric plasma–sprayed La 0.8 Sr 0.2 Ga 0.8 Mg 0.2 O 3 electrolyte membranes for intermediate–temperature solid oxide fuel cells. *Journal of Materials Chemistry* A 3: 7535–7553. DOI: 10.1039/C5TA01203A. 71

Zhang Y, Wang H–y, Jiang W, Bogaerts A (2015e) Two–dimensional particle–in cell/Monte Carlo simulations of a packed–bed dielectric barrier discharge in air at atmospheric pressure. *New Journal of Physics* 17: 083056. DOI: 10.1088/1367-2630/17/8/083056. 13

Zhang Z, Shneider MN, Miles RB (2007) Coherent microwave Rayleigh scattering from resonance–enhanced multiphoton ionization in argon. *Physical Review Letters* 98: 265005. DOI: 10.1103/PhysRevLett.98.265005. 12

Zhang Z–S, Crocker M, Chen B–B, Bai Z–F, Wang X–K, Shi C (2015f) Pt–free, non–thermal plasma–assisted NO_x storage and reduction over $M/Ba/Al_2O_3$ (M= Mn, Fe, Co, Ni, Cu) catalysts. *Catalysis Today* 256: 115–123. DOI: 10.1016/j.cattod.2015.03.012. 86

Zhao J, Yang X (2003) Photocatalytic oxidation for indoor air purification: a literature review. *Building and Environment* 38: 645–654. DOI: 10.1016/S0360-1323(02)00212-3. 90

Zhao J, Zhang Y, Yan J, Zhao X, Xie J, Luo X, Peng J, Wang J, Meng L, Zeng Z (2019) Fiber–shaped electrochemical capacitors based on plasma–engraved graphene fibers with oxygen vacancies for alternating current line filtering performance. *ACS Applied Energy Materials* 2: 993–999. DOI: 10.1021/acsaem.8b02060. 99

Zhao N, Ge L, Huang Y, Wang Y, Wang Y, Lai H, Wang Y, Zhu Y, Zhang J (2020) Impact of cold plasma processing on quality parameters of packaged fermented vegetable (radish paocai) in comparison with pasteurization processing: Insight into safety and storage stability of products. *Innovative Food Science and Emerging Technologies* 60: 102300. DOI: 10.1016/j.ifset.2020.102300. 31

Zharikov S, Shiva S (2013) Platelet mitochondrial function: from regulation of thrombosis to biomarker of disease. *Biochemical Society Transactions* 41: 118–123. DOI: 10.1042/BST20120327. 53

Zheng H, Chen P, Chen Z (2021) Cold–atmospheric–plasma–induced skin wrinkle. *EPL (Europhysics Letters)* 133: 15001. DOI: 10.1209/0295-5075/133/15001. 6

Zheng Y–Y, Ma Y–T, Zhang J–Y, Xie X (2020) COVID–19 and the cardiovascular system. *Nature Reviews Cardiology* 17: 259–260. DOI: 10.1038/s41569-020-0360-5. 56

Zhou F, Yu T, Du R, Fan G, Liu Y, Liu Z, Xiang J, Wang Y, Song B, Gu X (2020a) Clinical course and risk factors for mortality of adult inpatients with COVID–19 in Wuhan, China: a retrospective cohort study. *The Lancet* 395: 1054–1062. DOI: 10.1016/S0140-6736(20)30566-3. 38

Zhou G, Lan H, Gao T, Xie H (2014) Influence of Ce/Cu ratio on the performance of ordered mesoporous CeCu composite oxide catalysts. *Chemical Engineering Journal* 246: 53–63. DOI: 10.1016/j.cej.2014.02.059. 87

Zhou R, Zhou R, Xian Y, Fang Z, Lu X, Bazaka K, Bogaerts A, Ostrikov KK (2020b) Plasma–enabled catalyst–free conversion of ethanol to hydrogen gas and carbon dots near room temperature. *Chemical Engineering Journal* 382: 122745. DOI: 10.1016/j.cej.2019.122745. 101

Zhou R, Zhou R, Zhang X, Zhuang J, Yang S, Bazaka K, Ostrikov KK (2016) Effects of atmospheric–pressure N_2, He, air, and O_2 microplasmas on mung bean seed germination and seedling growth. *Scientific Reports* 6: 32603. DOI: 10.1038/srep32603. 24, 25

Zhou X, Cai D, Xiao S, Ning M, Zhou R, Zhang S, Chen X, Ostrikov K, Dai X (2020c) InvivoPen: A novel plasma source for *in vivo* cancer treatment. *Journal of Cancer* 11: 2273. DOI: 10.7150/jca.38613. 44

Zhu X, Gao X, Qin R, Zeng Y, Qu R, Zheng C, Tu X (2015) Plasma–catalytic removal of formaldehyde over Cu–Ce catalysts in a dielectric barrier discharge reactor. *Applied Catalysis B: Environmental* 170: 293–300. DOI: 10.1016/j.apcatb.2015.01.032. 88

Zhu X, Liu S, Cai Y, Gao X, Zhou J, Zheng C, Tu X (2016) Post–plasma catalytic removal of methanol over Mn–Ce catalysts in an atmospheric dielectric barrier discharge. *Applied Catalysis B: Environmental* 183: 124–132. DOI: 10.1016/j.apcatb.2015.10.013. 92

Zhu X, Tu X, Chen M, Yang Y, Zheng C, Zhou J, Gao X (2017a) La0. 8M0. 2MnO3 (M= Ba, Ca, Ce, Mg and Sr) perovskite catalysts for plasma–catalytic oxidation of ethyl acetate. *Catalysis Communications* 92: 35–39. DOI: 10.1016/j.catcom.2016.12.013. 90

Zhu X, Zhang S, Yang Y, Zheng C, Zhou J, Gao X, Tu X (2017b) Enhanced performance for plasma–catalytic oxidation of ethyl acetate over $La_1–xCexCoO_3+ \delta$ catalysts. *Applied Catalysis B: Environmental* 213: 97–105. DOI: 10.1016/j.apcatb.2017.04.066. 90, 91

Zhu X–M, Pu Y–K (2007) A simple collisional–radiative model for low–pressure argon discharges. *Journal of Physics D: Applied Physics* 40: 2533. DOI: 10.1088/0022-3727/40/8/018. 12

Zhu Y, Li C, Cui H, Lin L (2020) Feasibility of cold plasma for the control of biofilms in food industry. *Trends in Food Science and Technology*. DOI: 10.1016/j.tifs.2020.03.001. 71

Ziuzina D, Misra N, Han L, Cullen P, Moiseev T, Mosnier J–P, Keener K, Gaston E, Vilaró I, Bourke P (2020) Investigation of a large gap cold plasma reactor for continuous in–package decontamination of fresh strawberries and spinach. *Innovative Food Science and Emerging Technologies* 59: 102229. DOI: 10.1016/j.ifset.2019.102229. 30

Ziuzina D, Patil S, Cullen P, Keener K, Bourke P (2013) Atmospheric cold plasma inactivation of E scherichia coli in liquid media inside a sealed package. *Journal of Applied Microbiology* 114: 778–787. DOI: 10.1111/jam.12087. 24

Ziuzina D, Patil S, Cullen PJ, Keener K, Bourke P (2014) Atmospheric cold plasma inactivation of Escherichia coli, Salmonella enterica serovar Typhimurium and Listeria monocytogenes inoculated on fresh produce. *Food Microbiology* 42: 109–116. DOI: 10.1016/j.fm.2014.02.007. 36

Authors' Biographies

Dr. Zhitong Chen is a Professor at National Innovation Center for Advanced Medical Devices, China. He did his postdoctoral training in the Plasma and Space Propulsion Lab of Dr. Richard E. Wirz at the University of California, Los Angeles. He received his Ph.D. in Mechanical and Aerospace Engineering from the George Washington University, where he worked under the supervision of Dr. Michael Keidar. He received his M.Sc. and B.Sc. in Engineering Mechanics at the Institute of Mechanics, Chinese Academy of Sciences, and Mechanical Engineering at Northeast Agriculture University, respectively. His research interests focus on plasma physics, plasma applications, and medical devices.

Dr. Richard E. Wirz is a Professor in the Mechanical and Aerospace Engineering Department at the University of California, Los Angeles (UCLA), holds a joint faculty affiliate appointment at NASA JPL, and is an advisor to the United States Air Force and Space Force. He is the Director of the UCLA Plasma and Space Propulsion Laboratory and Energy Innovation Laboratory. He received his Ph.D. from the California Institute of Technology (Caltech). His research interests focus on plasmas, materials, and space electric propulsion (EP). He developed the world's first miniature noble gas ion thruster (MiXI) which is an enabling technology for precision formation flying of large multi-spacecraft interferometers, and critically enhancing for a wide range of new missions. His lab has also developed the first self–consistent, multidimensional hybrid plasma discharge model which has been used successfully in the design of ion thrusters of all sizes. His research in the area of plasma-material interactions (PMI) has lead to the discovery of the "plasma-infusion" properties of plasma-facing complex surfaces, which is a new material regime for high-energy density applications from fusion energy to propulsion. His Energy Innovation Laboratory has developed new approaches to wind and solar energy including the "biplane" wind turbine and sulfur-based thermal energy storage. Most recently, his lab investigated the effect of cold atmospheric plasma (CAP) on cancer therapy, especially for cancer immunotherapy, and virus disinfection. He has authored over 200 publications, 2 NASA Tech Briefs, and has several patents in these areas. He has received the AFOSR Young Investigator Award and the Northrop Grumman Excellence in Teaching Award.

Printed in the United States
by Baker & Taylor Publisher Services